The Foundations
OF
Engineering Contracts

The Foundations
OF
Engineering Contracts

M. O'C. Horgan
OBE, TD, DL, MSc, C.Eng, FIEE
and
F. R. Roulston
DLC, C.Eng, MIEE

Spon Press
an imprint of Taylor & Francis
LONDON AND NEW YORK

First published in 1989 by
E. & F. N. Spon Ltd
2 Park Square, Milton Park, Abingdon, Oxon OX14 4RN
52 Vanderbilt Avenue, New York, NY 10017, USA

First issued in paperback 2020

E & FN Spon is an imprint of the Taylor & Francis Group, an informa business

Typeset in 10/12pt Garth Graphic by
Scarborough Typesetting Services

ISBN 13: 978-0-367-58020-9 (pbk)
ISBN 13: 978-0-419-14940-8 (hbk)

British Library Cataloguing in Publication Data
Horgan, M. O'C (Michael O'Connor), *1910–*
The foundations of engineering contracts
1. England. Civil engineering contracts
I. Title II. Roulston, F. R. (Frederick R.), *1927–*
III. Horgan, M. O'C (Michael O'Connor), *1910–*
624

ISBN 0 419 14940 6

Library of Congress Cataloging in Publication Data
Horgan, M. O'C (Michael O'Connor), 1910–
The foundations of engineering contracts/M. O'C Horgan and F. R. Roulston.
p. cm.
Bibliography: p.
Includes index.
ISBN 0 419 14940 6 (U.S.)
1. Engineering contracts. 2. Industrial project management.
I. Roulston, F. R. (Frederick R.). 1927–. II. Title.
TA180.H67 1989
658.4'04'02462-dc20

Contents

Introduction

This is the first of a set of three books covering the needs of an engineer faced with responsibility for the management of a contract (or possibly a project, comprising a number of related contracts with different contractors). We have limited ourselves to engineering, using the word in its widest sense. We exclude administrative or financial reorganizations, such as stock-holding policies or methods of accountancy, though these may appear just as important to an employer (in his personal or corporate view) as purely technical or production ones. Our definition of 'engineering' embraces the design, construction or servicing of works be they civil, building, electrical, electronic or mechanical in flavour. A manager has been appointed to look after everything for the employer, or if the employer is a corporate body, his Board of Directors and their managing director. The project manager appointed will probably be an engineer, both by training and experience. Indeed, in some cases we shall examine, he is introduced – and his duties are expressed – by a Standard Form of Contract, probably as 'the Engineer'. His functions under such a title differ only marginally from what we expect him to do as a plain 'project manager'.

Probably for the first time in his career, a project manager is brought face to face with the realization that there is much more in his appointment than his technical expertise covers. He must embrace the equally important matters of commercialism, costs, passage of time and the law. They can all cost his client a lot of money, and he must avoid the pitfalls just as much as the purely technical aspects. As far as we know, there are no crash courses or polytechnic programmes by which he might widen his knowledge sufficiently in the time available to him: he must learn the hard way by bitter experience, making mistakes, and by leaving undone those things which he ought to have done. He leaves his client with a lot of unexpected expense and his directors with some consternation, together with his own lingering doubts as to his abilities.

This, then, is the need that we are trying to meet with the three volumes already mentioned – our own hard-won ideas are turned to ease the newcomer's burden, as based on our own experiences. We have not aimed what we have written at legal or contractual specialists (even though they might choose to use our efforts as sources of ready reference). Rather we have aimed to reach those newly concerned with handling contract management, and new contractors, in the flesh. At the same time, we have tried to make the contents pleasantly readable, and we have aimed at shortening or summarizing any legal statements. The latter (like all legal matters) ought to be presented in full, if only to ensure that correct interpretations are deduced, and it is to be hoped that our summaries will not mislead the reader. Neither of us is qualified or fully experienced in the law, so we have used only practical expositions to show how the law affects engineering contracts.

We make no attempt to train the individual in his engineering functions. It would be both presumptuous and impractical, when one thinks of the wide interpretation we have given to 'engineering'. We assume that having reached the dizzy position of manager, our engineer has already given full satisfaction to his superiors in his technical role. His selection might have been made by his client from:

- The firm's own engineering staff; for a full-time job he must be released from his normal duties entirely for the period of his secondment.
- An outside organization, such as a competitor, where he has gained sufficient knowledge of the trade to warrant his change of allegiance.
- A firm of consulting engineers specializing in project management matters; such a firm might already have been engaged to assist with the design of the present project, and this would be invaluable when it comes to execution.

We have divided our project into three parts, and allotted one volume of our work to each:

Volume 1 (the present book): *The Foundations of Engineering Contracts*
In this first volume we consider the many matters which affect a contract as a whole, including the numerous terms used in the later two volumes. By its very nature, it has a pronounced legal flavour, but it is an essential introduction to the first principles of project operation and should be mastered.

Volume 2: Competitive Tendering for Engineering Contracts This volume deals with the proper procedures for enquiring, tendering, appraisal and selection, so that a reliable contractor can be recommended by the project manager to his Employer.

Volume 3: Project Control of Engineering Contracts This volume covers the execution of the project in its methods, design and building on-site. The Employer elects to go ahead largely on the advice of his manager, who must now ensure his predictions are fulfilled. The Employer must not face unforeseen delays or extra costs at the last minute.

Thus an unversed manager is taken through all his duties, point by point, over the whole management period. He has to watch the development of the scheme and ensure that the Employer gets what he wants, functioning in the way he wants it, at a price he expects to pay, and by the date he has been led to expect it to be ready for him to use. Managers will – at their peril – overlook or play down a single duty, whether technical, legal or commercial, and to this end, they must have all details and their effects at their fingertips. It is only in this way that they can hope to appreciate instantly the significance of some minor occurrence on the prearranged programme. The time for detection and rectification of errors is at the start, before hours have been wastefully consumed and money frittered away on fruitless activities. Putting things right as a later exercise is always expensive for the Employer (who eventually has to pay for all wasted effort of this sort). Incidentally, he is also paying for his project manager for protection from such extravagances!

Possibly it might seem something of an irrelevance to bring it in at this stage, but actually it is not, when we are making the point that every small phrase matters. Nothing surpasses the surprise of, say, a young army officer instructing his raw recruits in the elements of map-reading. They suddenly appreciate for the first time that it is not just a matter of following a red line across a sheet; every single mark they come across (except possibly the last fellow's tea-stains!) has a meaning and a significance which they must learn to interpret and put to use. Contract documentation is much the same. Phrases have not been introduced in contract documents just because the author was handy at composition; nothing has been put on a drawing because the draughtsman was young and over-enthusiastic. Each phrase has a meaning and it is up to the project manager to find out what, and make full use of the information it gives. Complex sets of words (instead of the ones commonly used) and long, unbroken sentences devoid of punctuation were purposely chosen to give increased precision and avoid any chance of ambiguity. Whosoever may have occasion to read them later, long after the compiler has become 'unavailable', must be accurately affected in the desired way and in no other. A project manager is no exception.

A project manager is employed as a man of experience to do a specialist job, and those who deal in specialist advice cannot afford to make mistakes. He must know the sum total of his responsibilities from the start of his engagement; his future and his reputation will only be safeguarded if he satisfies the firm employing him. As they are paying him for his services, they expect him to do better than they might themselves. Extra expenses suffered as a result of a faulty contract can be both heavy and onerous, and can exceed by many times the full cost of the original project. The firm's scapegoat is ready to hand, and their manager will be blamed when they have to face such expense. We can only hope that our efforts will make the manager's task easier to assimilate and to fulfil. An early form of the present volume has been available in the guise of a short reference book bearing a similar title, but it has been completely rewritten and enlarged to bring it into line with the other two books of this series. They are also now available, so that the whole course of the project and its separate contracts have been covered and is ready for the use of budding project managements. References and appendices affecting matters dealt with in one of the volumes may be found in one or both of the other two, where they are more appropriately located: a project is a continuous undertaking and cannot be strictly divided into any three convenient parts.

Prologue

Extracts from Shakespeare's

The Merchant of Venice

(Act IV, scene I: A Court of Justice)

Shylock: And by our holy Sabbath have I sworn
to have the due and forfeit of my bond . . .
If you deny me, a fie upon your law!
I stand for judgement . . . shall I have it?
I charge you by the law, proceed to judgement.

Portia: Are there balance here to weigh the flesh? . . .
Have some surgeon, Shylock, on your charge
to stop his wounds lest he do bleed to death.

Shylock: Is it so nominated in the bond? . . .
I cannot find it; 'tis not in the bond.

Portia: A pound of that same merchant's flesh is thine;
The court awards it and the law doth give it . . .
Tarry a little . . . there is something else . . .
This bond doth give thee here no jot of blood;
The words expressely are 'a pound of flesh'.
Take then thy bond, take thou thy pound of flesh.
But, in the cutting, if thou dost shed
one drop of Christian blood, thy lands and goods
are, by the laws of Venice, confiscate.

Shylock: Is that the law?

Portia: Thyself shall see the Act.

Shylock: . . . Pay the bond and let the Christian go . . .

Portia: The Jew shall have all justice. Soft! No haste!
He shall have nothing but the penalty.
Therefore prepare thou to cut off the flesh.
Shed thou no blood; nor cut thou less nor more

> but just a pound of flesh. If thou tak'st more
> or less than just a pound – be it but so much
> as makes it light or heavy in the substance
> by the division of the twentieth part of
> one poor scruple; nay, if the scale do turn
> but in the estimation of a hair – thou
> diest, and all thy goods are confiscate . . .
> He hath refused his principal in open court.
> He shall have merely justice and his bond.
> Tarry Jew,
> The law hath yet another hold on you.
> If it be proved against an alien . . .
> He seek the life of any citizen . . .
> . . . the offender's life lies in the mercy
> of the Duke only, 'gainst all other voice . . .

And so on.

Motto for project managers: Know the basics of the law as it affects contracts (or bonds!) and get the wording right: no ambiguity; be precise. Say what you mean, and mean what you say. If you are a woman, there is no need these days to wear men's clothes, nor to change your name!

The White Rabbit put on his spectacles. 'Where shall I begin, please Your Majesty,' he asked. 'Begin at the beginning,' the King said gravely, 'and go on until you come to the end. Then stop!'

Alice in Wonderland – Lewis Carroll

1 *Engineering contracts*

1.1 THE PROJECT AND ITS ENGINEERING CONTRACTS

The contract is the bedrock of any engineering scheme – that is, apart from minor works which might be undertaken by one's own labour-force. At law, a simple contract (under signature of the parties concerned) is essentially a *bargain* between the two (or more) parties who have agreed to it. They could equally be organizations duly represented by nominated persons or others qualified to commit their organizations. The essentials of all contracts are:

- Something of value changes hands in each direction – the 'consideration' – which forms the 'bargain'. An undertaking or a gift by one party may be just as legally enforceable, but strictly speaking it is not a contract.
- The parties are in full and total agreement on everything the contract says, not just with the aims of the contract. Usually this agreement is manifested by one of the parties making an 'offer' and the other accepting it without changes or conditions.
- Any negotiations between the parties are on the terms comprising the 'offer', and take place before the offer (or a revised offer) is made. Subsequent negotiations may lead to a 'counter-offer', not an acceptance.
- Both parties want the terms of their agreement to be legally enforceable. Two friends may agree to meet for dinner, to split the cost, but they will not expect to go to court if one of them doesn't turn up – i.e. they do not form a contract as there is no desire to make their agreement legally enforceable.

There are, of course, other features which a proper contract must show, and we will be examining these in Chapters 2 and 3. For example, the persons signing an engineering contract on behalf of their respective companies must be legally empowered to commit them.

A storeman, a director's secretary or an office-boy deputed to sign the mail will not suffice. The company could later repudiate them and their contract, if it wanted to do so on discovering disadvantages in keeping to its undertakings. Some legal aspects of a contract are embodied in Acts of Parliament, by the words they use or the requirements they specify. Such a form is known as statute law, and to avoid or contravene it is to break the law. Other matters may have become common usage by tradition or the results of earlier cases heard by the judiciary. Such findings have built up what is known as the common law, which has been established as good and proper by earlier case law. In latter-day litigation involving any such feature of a contract, the party called to account must show why earlier dicta by the courts have not been obeyed, or how closely and clearly the present-day case resembles those of earlier cases. Such arguments are usually given by professional lawyers, who have the facilities and experience needed to locate and elaborate earlier findings. Matters are normally not left to a project manager alone to handle, and lawyers co-operate to reach such decisions. Nevertheless, he must be wise to the situation when he is dealing with tenderers, and appreciate when he needs to call for aid from the Employer's legal department.

Of the three volumes comprising the present work, this first is admittedly the most difficult for an engineer to read and to assimilate. It deals with the legal aspects, and however basically these are touched upon here, we must depart from the practical world in which the engineer has grown up and enter the more recondite ways of the legal profession. Contracts are essentially legal documents: it is necessary for any engineer to understand their arguments, and to follow their phraseology when evaluating any contract, or when setting out to prepare one. As far as possible, we have avoided the complex language lawyers habitually use in their striving for greater precision, and the special meanings they attribute to every-day words, together with their legal jargon and their not-infrequent use of Latin phrases. The last named can convey little to those of us who never got beyond 'amo, amas, amat . . .' or perhaps on special occasions 'dona nobis pacem'. It has been our endeavour to avoid misleading anyone whenever we have found it necessary to simplify the legal outlook. A first-aid manual never tells the whole medical story, it aims to give the person using it all he wants to know at a given time and place where first-aid is needed. He is enabled to take the right decisions and the correct action. In complicated or esoteric cases, the professional medical man is consulted and his advice obtained. A full diagnosis will be made and the person is told what he must do for the best. When such a point is reached, it is important in both the medical and contractual cases that nothing has been said or done at an earlier stage which adversely affects the treatment deemed appropriate by the professionals. A project manager must call for advice as soon as he finds a situation he cannot resolve on his own. On many occasions, in practice, the matter will work the other way round: the Employer's legal advisers may be keeping a watching brief on the progress of the project, and realize when their non-legal colleague is running into trouble. For example, the Employer will certainly refer to his legal advisers any contract he is advised to sign by his project manager. Hopefully their advice will support that of the project manager, and back his decision and applaud the steps he has taken to get the legalities right; but if he has overlooked important matters or made any grave mistakes, their attentions will act as a long-stop. They are on his side and will support him as far as they possibly can.

It must be underlined at this point that all contracts forming a project are between the *Employer* and his *Contractor*. However important he sees himself, and however much responsibility the contract gives to him for managing it, the project manager (or his equivalent, the Engineer) is *not* a party to the contract. His duties are strictly limited to those set out for him in the contract's terms, and consequently approved and accepted by the contractor. If he tries to go further, a contractor is entitled to remind him he has no right to do so, and can ignore his orders accordingly. A project manager must not issue any documents which accept a contract or alter its terms: he is not a party to the agreed bargain and cannot interfere. The Employer must do it for himself.

A project manager (even if he is the appointed agent of the Employer) must be careful that he does not unwittingly undertake responsibilities such as paying the contract price to the contractor. When he is called upon to represent the Employer as his agent, he must make sure that his actions are clearly specified and also that he is acting solely as agent on behalf of the Employer. He must (to protect himself from later accusations) make sure the appointment as agent states precisely how far he is expected to commit the Employer – and he must beware of going any further. The Employer 'owns' the project: his project manager is his employee to manage the contract on his behalf, and to offer him advice and what action he is recommended to take. Both the decision and the action are by the Employer himself, and a project manager must take all steps to protect himself if he is ever called upon to do the Employer's work for him.

Luckily, as we shall discover later, most of the Standard Conditions of Contract allot space for the naming of the Employer's manager (or the Engineer) and their subsequent clauses spell out his duties. These have already been accepted by the contractor when signing the contract.

1.2 HOW CONTRACTS ARE FORMED

Contrary to popular belief, contracts do not all have to be in the form of comprehensive documents. A contract can be made verbally or on a scrap of paper, the back of an envelope, always provided that it fulfils the necessary legal requirements. Many of our daily activities, which we undertake without giving them a second thought, are really perfectly good legal contracts and we can be sued if we fail to fulfil the promises that we have undertaken, for example, ordering a meal at a restaurant and then not paying for it. A contract may arise in one of several different ways.

1.2.1 Contracts formed solely by actions

Such a contract may involve no paper-work or any specific verbal agreement; the example of the restaurant, above, is a good one. The waiter proffers you a menu, making an offer on behalf of the proprietor at the price stated therein. There is an implied term that you will pay for whatever you order as soon as you have consumed it, before leaving the premises or at some other time by prior arrangement with the proprietor or his agent. You make your choice of dish and order it, it is subsequently brought to you and you eat it. Your acceptance of the offer is implicit in your accepting the proffered dish and consuming it; no further acceptance is necessary to complete the every-day 'offer and acceptance' formula. You then have to carry out your part under the contract with the proprietor – you must pay the sum arranged.

A purchase at an auction is an interesting example of a contract formed by action only, and we shall refer to it below in another context. You make a bid (offer) by waving a paper or twitching your ear, from the body of the crowd. In due course, the auctioneer accepts one of the offers (yours) by banging his desk with his gavel. You have yourself a contract and must pay the sum you offered (bid), but no word has been uttered. There are implicit terms in keeping with the universal customs of the trade of auctioneering, and any court would enforce them if appealed to by either party to the purchase.

At first sight, one might be tempted to think that the same holds good for a railway journey. British Rail offer to take you to, say, London for a stated sum of money. You duly buy your ticket from their appointed agent (the booking-clerk) and the train is presented to you at the platform. You board it and take the trip, hopefully being duly delivered to London. However, on examination, this contract is not based solely on actions: if you look at the reverse of your ticket, you will see that British Rail have drawn up a full contract in writing and you are deemed to have read it and accepted its terms from your action in buying a ticket. Copies are available at every booking-office, so that you can actually read the clauses first. If you do not, that is simply your fault, not theirs; the conditions of contract were drawn up by British Rail, so they may be said to favour the rail people and not the traveller. They form an intrinsic part of their 'offer' to transport you to London. They disclaim all responsibility for being late or for you having to stand all the way in a crowded train, even for not getting you to London at all if they are prevented by something outside their control. Here the technical term is 'frustration' of the contract. You have accepted it all! It would be nice (if one had the money!) to challenge them for such failure, on the grounds of 'negligence' and breach of contract, especially having in mind their latest obligations under the more recent statutes. If they go unchallenged, much internal bungling can be laid at the door of 'Force Majeure'.

A project manager must keep his eyes open. He may find himself faced with a stranger who has a lorry-load of brick hardcore from a demolished building, or so he says. He offers it for a 'couple of quid the lot', and the project manager tells his clerk of works to dump it into his stock, pay the lorry-driver and get rid of him. By accepting delivery and making payment, the project manager has accepted the offer and formed a contract with the lorry-driver to buy the contents of the lorry. Has he considered his position if:

- the lorry-driver does not, in fact, own the hardcore, but it belongs to somebody else?
- the top surface is good, clean hardcore, but underneath it is dust?
- the lorry-driver sends him a bill for transporting and dumping it at the site?
- the lorry is only half-full – i.e. a half-lorry load?
- it is not clean hardcore, but recovered from a dump where it has lain for years and accumulated a lot of soft earth during recovery?

This leads on direct to the verbal contract.

1.2.2 Contracts formed verbally

Legally, this approach could be used for most engineering contracts, other than those involving land (see below), but the trouble arises later in establishing just what has been agreed. Witnesses never listen to details of what is discussed, or appreciate the points likely to prove important, if some disagreement arises, and the case inevitably comes to the courts for settlement. A wealthy relative in the Midlands (at the time when there was a universally recognized statement that 'an Englishman's word is his bond') transacted all his business in the way the Midlands knew, with some notes on a scrap of paper (unsigned) and a shake of the hand. When he died, his solicitors spent over three years trying to establish his exact position, and even so failed to satisfy the probate authorities. His executors had a harrowing job. He was found to own brickworks, collieries, housing estates, and goodness-knows-what else, all bought from their previous owners with nothing more than a cheque and a handshake.

The writing is thus on the wall for all sales and technical staffs, striving to complete a deal with a customer. They must not postulate anything verbally that they are not in a position to carry out. They may well find themselves 'stuck' with some technical improvement or elaborated specification – much-improved example of a modern scientist's dream – but at the old price! The writing is on the wall for the project manager, too: he must insist on everything being recorded in agreed minutes of meetings, in letters, memoranda and, especially, in contract documents. We have already made the point elsewhere, but we repeat it here with emphasis: *have it recorded in writing.*

1.2.3 Contracts required by law to be in writing

There are some subjects which are required by law to be dealt with by contracts in writing – i.e. the whole contract must be written out. They include contracts dealing with the transfer of stocks and shares, bills of exchange, promissory notes (notes beginning 'I promise to pay . . .') and marine insurance. Some purchase contracts using the special rules for hire-purchase under the 1965 Act also need to be contracted in writing. Generally speaking, none of these subjects is likely to come the way of a project manager stationed in the UK: any such matters would be dealt with by the specialized professional staff who are employed by his Employer. He may meet them, however, when he is at a site abroad and acting with additional powers from his Employer because of large distances involved. Largely for this reason, we have included this class of subjects, so that such a manager will be aware of their special needs.

1.2.4 Contracts required by law to be 'evidenced in writing'

There is a significant distinction between a contract which is required to be 'evidenced in writing' and one which has to be completely in writing. The former phrase calls for something less formal, namely that it must be backed up by memoranda which have to be signed by the parties and retained as available evidence. They must between them record all the material terms of the transaction: the names and addresses of the parties, the subject of the transaction, the consideration, terms of payment, contractual dates, and so on. The memoranda should preferably be signed by both parties, but one will suffice as long as it is clear what it purports to be. The memoranda need not be specially prepared, but can comprise several letters, minutes of meetings, records of negotiations, etc. as long as all relevant details are covered. One such subject relates to money lending and this is not likely to affect our project manager; others which might affect him include:

- Contracts relating to land.
- Contracts relating to guarantees.

The former must include a definition of the land, and whatever the phrase is intended to cover (not only its location and extent, but also any specified growing crops, growing and felled timber singly or in plantations, specified and described buildings or machinery erected on or lying on the land as described). The record must also apply to the use of land – e.g. leasing; disposition of rights-of-way; rights of shooting, fishing, hunting or boating and nowadays water-skiing; rights of bathing; removal of soil or mineral ores from the land; taking away water or soil from the land; movement of vehicles over the land, free-ranging or by tracks already existing; and so on. Any of these might affect a manager in charge of a site, especially one who is remote from his Employer and having to stand on his own.

The second subject of guarantees should not be confused with their near-relatives, warranties, especially in the sense of an undertaking of freedom from faults. A 'warranty' is also a class of term of contract (we deal with this more fully under breaches of contract). It is one of the features of a guarantee that is given by a *third party*, that is one of the parties is 'backed' by a third party who stands as his Surety or Guarantor.

A party to a contract cannot, himself, give a *guarantee* – it then becomes a 'warranty', and the party is said to 'warrant' such-and-such. The *third party* undertakes to make a specified restitution (usually in the form of a sum of money – up to a stated maximum amount) if the party he is guaranteeing *tries but fails* to fulfil his obligations under the contract. Note that a guarantee is not an alternative solution which a party may choose to adopt: the second party must have tried and failed before a surety will accept action. He will rarely guarantee 'specific performance' (Section 5.5) but normally restricts himself to a sum of money. The subject of guarantees is a complicated one, and we return to it in more detail in Chapter 5.

1.2.5 Written contracts

This is the form of contract normally negotiated between two parties interested in a project, and the one most frequently met with or adopted by the project manager. It closely resembles the class mentioned above in Section 1.2.3, except that it is at the discretion of the parties, and not required by the laws of the land. A written contract can take several forms but, in each case, the clauses and conditions which express the intended courses of action to be taken in various situations are spelled out in legal terms, and these are either included in the contract itself or referred to in the documentation.

In its simplest form, a contract may comprise a formal order (on one of the purchaser's standard order forms) accompanied by his standard conditions of purchase. It is accompanied by a simple acceptance slip, which the seller signs and returns. Note that such an order and acceptance together form the purchase contract, and supersede any earlier negotiations between the two parties regarding the purchase. As the order form is sent out by the Employer's purchasing department, it may be done without the knowledge of the person who has been actively negotiating, and may indeed upset the arrangements made during the negotiations regarding the purchasing deal. This is a most fruitful source of error – and it should be looked for carefully where a standard form of order is used to 'confirm' the purchase of anything by the Employer. In its more complicated form, a written contract can comprise a full document, either under hand (i.e. signed) or under the corporate seal of the Employer and the Contractor (see Section 5.2); it will include the conditions of contract. We shall look more closely at such contracts later in this first volume in Chapter 5. A written contract will be the normal one for all engineering works, and again we shall deal with it in some detail, including Agreements, Bonds and Guarantees. We look at Standard Conditions of Contract in Appendices 6 and 7.

For most engineering contracts, especially if any work is to be carried out on the Employer's site, the printed business contract form used by any party is usually inadequate, and if used, it will introduce numerous new considerations. Luckily for the project manager and his team, it is not necessary to dream up new terms for these on every occasion as most industries, large manufacturers, institutions, public bodies and similar large dealers have developed over the years model sets of standard conditions of contract which they have found readily applicable to their business.

These 'Model' or 'Standard' conditions of contract supply most of the necessary backing, and they are available to all and sundry in printed form. They can be incorporated into any contract documentation by a project manager or a contract manager, by direct reference to the formal title of the set of terms chosen. If this selected standard does not fit exactly the circumstances he faces, a manager can devise and add supplementary clauses or modify the standard ones to make good the deficiency. Usually he will recognize from the outset in forming or negotiating a contract the particular standard sets of conditions which are preferred by the firms to whom he intends sending his enquiry, and the consequent points of difference can be negotiated at the discussion stage, before a formal offer is made.

The addition of supplementary clauses is preferable to the amendment of existing ones in a set of Standard Conditions of Contract; the latter are usually carefully interwoven one with another, so that any modification in one place often leads to consequent modifications elsewhere in the document. This succession of changes can be avoided by a supplementary clause which might start, 'Notwithstanding anything to the contrary elsewhere in this document, the such-and-such will be dealt with as follows . . .'. Such an addition automatically overcomes any succession of clauses which otherwise might easily be missed. (We shall have more to say about standard sets of conditions later in this volume.)

Whenever a document is referred to and clearly specified in a contract itself, it automatically becomes a part of the contract documentation, even if it is not affixed to or accompanying the contract itself. Thus, for example, a specified insurance policy, Model Conditions of Contract (stating the year of issue thereof!), British Standard Specifications (identified by the serial number concerned), sets of Employer's drawings, and the like, can all be called up by reference and become a part of the documentation affecting the contract. A great problem is deciphering exactly what has been undertaken or promised at the time a verbal contract is made, and what additional standards or drawings have been referred to (especially when the verbal contract was made several years previously). Witnesses are vital, and their views should be recorded at the time. Memories on important details or actual wording are notoriously bad, more often than not things have to be left to a decision by the courts from the other relevant evidence as to just what the parties to the verbal contract intended at the time that they made their bargain. Sometimes it will be found that previous contracts were not in force at that time or witnesses recollect matters discussed at the negotiating period which were later dropped, or that necessary matters (such as 'consideration') were omitted or, indeed, that the supposed contract never existed at all! Even local customs and traditions may be shown to be unknown to one of the parties concerned. We have already stressed the course that is open to a project manager – namely, to get everything written down and agreed to by both parties at the time.

1.2.6 Standard order forms

We have given a word of warning in Section 1.2.5 regarding a habit which many firms have of issuing one of their standard order forms (probably because of its order number), to cover the acquisition of the contract works, and so introducing it formally into their accounting routine. It often happens that a collection of their senior officials has striven for many months with one of their customers, to negotiate a contract dealing with these works. This they have finally achieved, and the formal order from their bureaucratic brethren is the result.

Unfortunately, this order form has printed on it, usually in the fine print, a sentence which runs, 'This order is placed subject to our normal conditions of sale as follows . . .'. An acceptance of the order (usually sent out of politeness, if nothing else) replaces the earlier contract (as negotiated) by the 'normal conditions of sale', the order plus its acceptance tcgether forming a second agreement, neutralizing the first. No doubt, the position (if and when it is realized) can eventually be sorted out, but the simplest approach to prevent all present and future difficulties is for the acknowledgement of receipt to be specifically that and nothing more. It neither accepts the order nor replaces the hardly-achieved contract. The firm is requested to withdraw all mention of 'normal conditions of sale' and replace it by the recently negotiated terms of the contract.

A copy of this letter to the senior member of the firm who was their leading negotiator will, of course, ensure that its recommendations are accepted and followed up. The purely administrative measure no longer takes control, and the contract as negotiated remains valid.

1.3 EXPRESS AND IMPLIED TERMS OF A CONTRACT

A contract may be made up of written words, spoken words or conduct, or a mixture of these. Any words (whether written or proved to have been spoken) which the parties are agreed shall form part of the contract (or the whole) are known as the *express terms*. However, there may be other terms which naturally follow as part of a contract, even though they may not have been uttered: these are the *implied terms*. In this section we deal with these two expressions and the terms to which they are applied.

1.3.1 Express terms

Where a contract is wholly in writing, its terms may be either in the contract itself, or in some other document which is referred to and identified, and hence automatically becomes a part of the contract documents. It may still be held by a court of law that an unidentified document must be considered to be a part of a contract, but this is solely a matter for the court, and the document concerned must be one which affects the contract as it stands and ought properly to have been introduced by a suitable reference. A problem arises when a contract is partly or wholly a spoken (verbal) contract, and the main task of the court will then be to determine exactly what words were used and what they meant.

There is a similar problem if a contract is partly written and partly spoken. It must first be decided whether the spoken words did indeed form a part of the contract at all, or whether the written part comprises the whole of the binding contract between the parties. There is a general rule that a spoken word must not add to, vary or contradict anything that has been written and signed (on the ground that the written agreement would have been altered accordingly if it had been intended to amend it by the spoken words). Applied strictly, this rule can lead to injustice, and it is not therefore applied to matters which might affect the validity or the period of operation of the written contract. Yet another exception occurs where it is clear from the other evidence that the written contract does not form the *whole* of the agreement between the parties, and both parties agree to the spoken terms being included.

The court must always be mindful, in considering additional evidence, that various terms might have been proposed and discussed during the negotiation phase of a contract, then abandoned when it came to drawing up the final contract. Thus evidence that such measures were considered is, of itself, no guarantee that the parties both agreed to them being part of their contractual obligation.

The general rule of inadmissibility of spoken words over the written word does not refer to evidence of local traditions and customs where it can be shown that both parties were aware of the existence of such customs and traditions and should have taken cognizance of them, even through they were not expressely mentioned in the written contract.

1.3.2 Implied terms

The terms written into a contract may not show everything, especially when every-day matters such as might be assumed by both parties are concerned. Sometimes disagreements may arise which hinge on a term which was not included, and it is then necessary for the correct interpretation to be introduced. In such cases, the disagreement can be taken to the courts to come to a decision as to how the missing term should be legally brought into an 'agreement' between the two parties; it is then referred to as an 'implied term'.

Before reaching any decision, the court may examine evidence well outside the contract itself, and embrace events during earlier negotiations or discussions leading up to the contract, in order to try to establish exactly what the parties had in mind at the time of signature. There are three types of implied terms:

1. Those which earlier laws (statutes) require the parties to introduce or at least to accept. The Sale of Goods Act 1893 amended by the Supply of Goods (Implied Terms) Act 1973 which was, in turn, amended by the Unfair Contract Terms Act 1977 is full of such examples. In such cases, the court has no problem in ordering them 'in'; indeed a good company lawyer would prevent such questions coming to court at all.
2. Terms required by conventional usage in a particular business or trade. The parties may expressly include, or expressly exclude, such terms – i.e. they have the option of not following the normal usages of the trade, but they must say so in the contract; otherwise a

court will consider that such a course was accepted, but being obvious to both parties was not included. The term is then included as an 'implied term'.

3. Those terms introduced by the courts themselves, as a result of their examination of witnesses and other evidence and after attempting to establish what the parties originally intended. Such implied terms must be:
   (i) reasonable;
   (ii) necessary in the circumstances;
   (iii) related to subject-matter known at the time of signing the contract to *both* parties – a decision regarding a matter quite unknown to one of the parties could never be justly incorporated into an *agreed* contract;
   (iv) rarely other than unsophisticated.

It is never the object of the courts to rewrite a contract or to assist one party to reword an unequal bargain, or one badly drafted. The parties have had the chance to read the document, to seek the advice of experts and to negotiate phraseology: they have made their agreement and signed it. They must abide by their decisions; someone who has negligently accepted a slovenly or badly drafted contract cannot be rescued. Therefore a term included by the court as an implied term will not upset any decision by the parties on a key matter or alter the contractual aims, or respective duties, of the parties who signed it. The court will restrict itself to every-day matters. In one court a judge referred to the situation as follows: 'if, when the contract was under negotiation, a bystander had said "What will happen if so-and-so occurs?" and both parties had replied with some vehemence, "Why, such-and-such of course . . . it's really so obvious that we haven't troubled to say anything about it".'

We have mentioned above those terms which are required to be followed in the light of earlier statutes. In the past it has not infrequently happened that a certain implied term has been found necessary on so many occasions that it has eventually been incorporated as a required term. Much of the Sales of Goods Act arose, in this way, from what had previously been 'implied terms'. For example, the requirement that when ordering from a sample, the items supplied would be the same as the sample; or that if an object is ordered to fulfil some purpose, it will be reasonably fit to perform that purpose. Quality must always be of a merchantable standard. Originally the parties could exclude such terms from their agreements if they so wished, but subsequent legislation has taken away this option. The Health and Safety at Work, etc.

Act 1974 is also largely of this character when it deals with conditions at a place of work.

It is most important when two parties have made a reference to the courts for a legal interpretation of a point in their contract, that the courts should make a firm ruling, one way or the other, and not 'sit on the fence' by saying that the evidence is equally balanced and both of them are right! Thus, in cases in which they find the evidence balanced, they must first resort to a number of rules by which their decision is determined. The most important of these are the following:

- Words are given their ordinary, every-day meaning and not an esoteric or abstruse one. Similarly, interpretations of phrases should be reasonable, in preference to more unreasonable ones.
- Odd phrases by traditional trade usage can be supported (e.g. a 'glass of beer' (traditionally about a half-pint), a 'yard of ale' (traditionally a tall glass-full, or a special glass-full), the height of a horse ('hands' from floor to withers), the cost of a pig (traditionally per 'score' of 20 lb, not per animal), a 'baker's dozen' (traditionally 13), and the like.
- Obvious errors or absurdities in a document must be corrected before a judgment is made.
- A special clause, written in by the parties, takes preference over a 'standard' model clause with which it is not consistent; it is at least 'thought out'.
- The omission of an item from a list (i.e. two or more) of items of one and the same type is considered to be left out on purpose; for example, if copper is omitted from a list of metals, it is interpreted as intentional. Similarly, if three items in a list were 'to be supplied and delivered', a fourth item, left unspecified, might be considered as to be supplied but not delivered.
- If the meaning of a word is doubtful, it may be assumed to belong to the same class as an associated word in the phrase. Thus, for example, 'other materials' as a sweep-up at the end of a list of metals could be assumed to mean 'other metals' rather than the indefinite 'materials' (sometimes referred to as the Ejusdem Generis Rule).
- The words of a document are often construed against the party which had hoped to benefit by them (often the same party which had introduced them into the contract, but not necessarily); for example, a vague clause dealing with 'extensions of time' could be construed against the party which otherwise might benefit by less liquidated damages to pay. This is often referred to as the Contra Proferentem Rule.
- The recital to a document, though strictly speaking not a legal part

of that document, may be invoked if it showed what the parties had intended at the time.

The courts may sometimes use the statutory terms from one Act when considering applications for guidelines in another. The actual 'implied terms' may be modified to suit the different circumstances, but arguments can be properly used across a borderline for consideration, even if they are not finally adopted in a recognizable form. We may have dwelt rather too long on matters of buying and selling, but it is only right to say that a project manager may well come up against the subject of 'implied terms' in contracts which have nothing to do with those activities. Do the conditions at his site fulfil the contracts of employment of many of his contractor's men? Are the requirements of the Health and Safety at Work, etc. Act up to standard? He and his contractor might easily run into strike trouble, with a consequent delay to his Employer's project if there are shortcomings associated with his project site.

As always, the Employer holds the project manager responsible, and he must recognize his duties, even though they might involve decisions outside his normal role. The latter may well occur at a site from which contract with the Employer may enforce a long delay to the work, in face of the implied contract requirement.

1.4 AVOIDANCE OF AMBIGUITY

Nothing could sum this up better than the following extract from *Alice through the Looking-Glass*, Chapter vi:

> 'When I use a word' Humpty-Dumpty said, 'it means just what *I* choose it to mean – neither more, nor less.'
> 'The question is,' said Alice, 'whether you CAN make words mean so many different things!' Humpty-Dumpty began again: 'They've a temper, some of them – particularly verbs; they're the proudest: adjectives you can do anything with.'

Clearly contracts, as legal documents enforceable at law, must be unambiguous. Everybody who has to read or interpret them must do so in the same way. A contract may continue in force for several years before everything is cleared up and the statute of limitations has ceased to be a factor. During this period many who drafted or negotiated a contract originally may have disappeared from the scene, but their intentions must still be accurately deduced in their absence. The only legacy they have left behind is the written word, so this must have but one meaning to their successors, their legal advisers, opponents' and arbitrators, as well as members of the judiciary, when they are called upon to deal with the affair. The wording not only needs to be unambiguous, but meanings attributed to individual words must be specific, especially if their definitions are not the ones found in standard dictionaries. Each contract should therefore be provided with its own glossary of terms defining any doubtful words. See Appendix 1 in this volume; in the examples we give typical ones in daily use – the definitions we attribute to them are also given. Where this differs from, or is more restricted than, that given in recognized dictionaries, the word is (by custom) spelt with an initial capital letter. Thus the 'Employer' (one of the parties to the contract) means more than a user of labour, or of something, and as such we use the word with a capital 'E' throughout this book: indeed at one point we have been forced to describe a body as being 'both the Employer and employer' of someone else! A Site is something legally different from a site, the Works are the contractual works, and a 'month' is the one according to a given type of calendar, of importance in countries which do not adopt the Gregorian calendar. A detailed reading or legal interpretation of a contract must always be preceded by a study of the limited interpretation of its words, as given in the glossary which should form a part of every contract. If it is not stated, the meanings to be attributed to its words may be the same as those used in a standard conditions of contract called up by the contract . . . or they may not! *Check them.*

Absence of ambiguity is the hallmark of a well-drafted legal document such as a contract. Not only is the form of words carefully chosen, but it will frequently appear at first sight to be unnecessarily repetitive; on deeper reflection, it will be realized that the description defines more closely what the writer intended. The choice of words and phraseology is necessary if the draughtsman is to be sure of including every ramification he requires in his final term. Thus he may 'own, have, receive, obtain, find or borrow' something, and these may all appear at first sight to imply much the same thing – by fair means or foul he has got his hands on the thing! But not so for the legal professional: 'own' means it belongs to him; 'have' means it is in his possession (but he doesn't necessarily 'own' it); and 'receive' indicates he does not have it at the time the contract is signed, but he may be given it afterwards and subsequently get his hands upon it. There is a matter of intent in what follows: 'obtain' suggests he does not have it, but will subsequently actively seek it; 'find' likewise suggests he subsequently comes across it by chance; and 'borrow' has its usual implication of getting his hands on it temporarily, ownership remaining with the lender.

In its way, each word means something slightly different; not perhaps the best example, but it shows why a group of apparently synonymous words is sometimes met with. We non-professional readers must bear with them and try to discover why it has been thought necessary to include them all! A case was experienced in which insufficient definition led to an extensive waste of time, money and argument over differing views of what the word 'system' (in the phrase 'electrical system') meant, what it included and what it did not.

It might seem hardly credible, but when a contract had 'gone bad', a company and its legal advisers negotiated with the second party and actually started to prepare a case for the courts, before it was pointed out to them that while they were using the current edition of a standard form of contract conditions, the actual contract called up a long-since dead edition, which had been out of date for a number of years; however, this was of course still valid for the contract which had used and identified it.

Both of the above examples illustrate loopholes in a contract glossary, and serve as a warning to project managers which is certainly not unjustified. Ambiguities *now* invariably lead to arguments *later*.

A government draftsman working on a statute made it apply to all 'shops and offices'. His colleague reminded him of market stalls, and women with baskets squatting at the roadside. So he wrote: '. . . shops or offices, and places used as shops or offices . . .' and believed that thereby he had got at the lot. But he hadn't: he had forgotten the stop-me-and-buy-one tricycle! The Lords decided this was neither a shop nor a place used as a shop, and the law did not apply, which serves to show just how far one must go to avoid ambiguity or doubt.

1.5 BREACH OF CONTRACT: CONDITIONS AND WARRANTIES

The terms ruling a contract may not all be of equal importance. Inevitably some will have a greater effect on the outcome of the bargain, and traditionally the courts recognize and use a more or less uniform terminology to divide them into two classes. These are known as 'Conditions' and 'Warranties', quite regardless of any other meanings we apply to the same words in dealing with engineering contracts. The names are not well chosen: they must not be confused with 'Conditions of Contract' (by which we mean the collection of *all* terms applying to the contract, both legal Conditions and Warranties), nor with the warranties given by a manufacturer to the Employer, being solemn assurances that his product will be of a stated standard of excellence and without any flaws. Or possibly, that a stated procedure will be faithfully pursued.

The legal division of a contract's terms when referring to breaches of contract is as follows:

- *Conditions.* These are the more important terms, which if not performed or obeyed by one of the parties would cause the other party to obtain from the contract a result basically different from that for which he originally contracted.
- *Warranties.* These are terms of lesser importance, the second party still obtaining the main purposes for which he contracted, but he would be faced with a minor difference for which he could be adequately recompensed by a payment of money by the offender.

The terms of any particular contract are not specified as one or the other: they can be so assessed by the parties, in the view of all the existing circumstances, or (in difficult cases) would be so attributed by a court. For a breach of a 'condition', the injured party is legally entitled to treat the whole contract as at an end (i.e. legally to repudiate it) and to claim from the offending party all damages he may suffer as a result, from having to start again with a new contractor. However, he may if he so chooses continue with the contract (i.e. reaffirm the contract) and still claim such damages as he may incur by having to rectify the breach and thereby being delayed. A project manager might well choose to use this alternative to save the longer delay if he has to start again from scratch with a new contractor.

For a breach of a warranty, the victim has no right to terminate the contract, but he can claim recompense for the differences with which the breach of warranty may have presented him, for example, a poorer product, reduced output from a plant, more expensive operation, and delay in taking a plant over and having the use of it. A case in point might be the contractual date for completion of a plant on-site. Where liquidated damages are specified in the contract, these are all that the purchaser can obtain for delays during that length of time. However, it has been held that where a maximum value is set on liquidated damages, the delay which leads to that figure is the maximum delay the parties are prepared to envisage. Anything longer can be taken as an unexpectedly long delay or even repudiation of the contract by the delaying party. The delay clause might be regarded as a 'warranty' during the liquidated damages period but a 'condition' during any *extended* delay. If it is to be clear to both parties from the outset, the fact that completion date is always a 'condition' may be underlined by the use in the contract of the phrase: 'is of the essence of the contract.' Such a phrase can be used in any connection (and not only with a completion date) where it is intended to make it clear that the breach is of a 'condition', with all that that entails.

Here it is perhaps necessary to give a warning connected with the use of the phrase that the payment of damages 'is in full satisfaction for any delay'. Such a term is sometimes met with, especially in connection with model sets of conditions of contract. It is dangerous, in that it puts no limits on the *length* of the delay, and precludes any additional penalty which might arise if the term were adjudged a legal 'condition', such as termination of the contract, however long the period of the delay might exceed the one for which liquidated damages are prescribed elsewhere in the contract.

The division of the terms of a contract as between legal conditions and legal warranties may not be easy. Each has to be taken on its own merits, and due regard paid to the actual words of the term, the effects of the breach, the speed of reaction of the other party and their actual actions taken as a result of the alleged breach. It is, of course, always necessary to evaluate the result of the breach, the situation it has produced, effects on the outlook of the injured party and the extent to which he can be recompensed purely by being given a sum of money. If a party contracted for 'sausages', did he get sausages but without skins, or did he finish up with liver pâté? If he wanted a concrete structure, did he get a brick-built one, half the size? If possible, the parties must decide the matter for themselves, but the results of a breach of a condition are so far-reaching that the matter often has to go to arbitration or to the law courts, the ultimate deciding authority. The essence is, of course, how far what the injured party gets meets the essentials of his needs – i.e. what he expected to get from his contract.

On a different subject, the Law Lords more recently confirmed the conception of another sort of term, even more fundamental than the condition. They have agreed the existence of what is known as a 'fundamental' breach, namely one which attributes to one of the parties a total failure to perform his part of the contract. This might be by sheer inability on his part, or by the provision of something quite different from what he contracted to produce: 'He asked for bread, but they gave him a stone.' It is not merely the breach of a condition, but of the whole purpose of the contract itself. The importance of this decision has been, to some extent, reduced in recent years by the increasing use made by Employers of the principles of Quality Assurance, a procedure we shall refer to later in this volume and elsewhere in the present series of works.

1.6 THE PRIVITY OF CONTRACTS

Although there is a so-called Doctrine of the Privity of Contracts, this is really nothing more than a glimpse of the obvious when we remember that a contract is a bargain between the parties to it, be they individuals or corporate bodies. They have made the contract, and they alone can alter it. The doctrine states, in effect, that a person who is not a party to a contract can neither enjoy its benefits nor assume its obligations. In other words, it must follow that such a person can neither sue nor be sued by anybody else under the terms of the contract and according to contract law. The person is, of course, still an individual, owing certain duties to his neighbours, and he may still be charged with dereliction of such duties by any other individual or person under common or criminal law, but not under the law of contracts. Thus he can be sued for a car accident, or stealing or any such offence, but not under the terms of a contract to which he is not a party.

The full implications of the doctrine are complex, and not a little obscure – especially when it comes to benefits. Thus agents, trusts, assignees, leaseholders and others may become involved if they are (or become) parties to the contract. It is not likely to affect a project manager (assuming he is not himself a party to the contract). Normally his appointment is made by the Employer (with whom he may have a separate contract!) and he cannot himself expect any benefit from a contractor under the terms of the works contract. For most project work, the project manager is not a party to the contract and he himself cannot sue the contractor, nor for that matter the Employer, unless he has such a separate contract of employment with the latter; it is also the latter (not the main contract) who pays his fees.

Usually the Employer, in his turn, has no contract with any of his contractor's subcontractors, and hence no right to sue them under his main contract. This applies whether the subcontractor is nominated or not, but in the latter case (as we shall see when we consider nominated subcontractors in greater detail), it is usually recommended that an Employer tries to set up a collateral agreement for this very reason with the subcontractor. He becomes very vulnerable if the nominated subcontractor is late with his work, or if some joint development fails to operate satisfactorily when installed under the subcontract. This is not entirely straightforward: an Employer cannot take action under just any collateral agreement. It must deal specifically with the actions to be pursued should a particular shortcoming in question crop up. Alternatively, the subcontractor has his usual rights in agreeing to a bargain direct with the Employer. He may not be agreeable to entering a collateral agreement at all, or he may require such a high consideration that the whole matter becomes uneconomic from the Employer's standpoint.

The servants of either party to a contract are not themselves parties to the bargain and cannot be sued. This applies not only to ineffective workmen, but to senior staff and to managers. The main contract with the Employer may require the contractor to remove them from the project, but any dealings with the individuals themselves must be taken by the contractor under his contracts of employment with them.

By and large, the Employer is in a somewhat vulnerable position. If his contractor is in delay or produces bad work, he can only sue him under the terms of the legal contract he has with him. He depends a lot on his project manager:

- Only recommending reliable and proved contractors.
- Only recommending foolproof contract documents, both contractually and technically.
- Having a rigid system of management which forecasts inaccuracies and delays.

Here the value of Quality Assurance Certificates becomes clearer. Normally speaking, the Employer has no control over the contractor's chosen subcontractors and can only get at them through his chosen contractor. It is largely for that reason that he will seek to approve a contractor's list of subcontractors at the enquiry stage of a contract.

1.7 COLLATERAL CONTRACTS

A collateral contract is a properly constituted contract running alongside and referring to a main contract. It may often deal with the relationship between one of the parties of the main contract and a third party, not a party to the main contract. It is an essential feature of collateral contracts that they extend or relate to the main contract to which they are 'collateral'. It is not sufficient that they are merely those contracts which run contemporaneously alongside another contract – e.g. both forming parts of the Employer's project. The term would not be applied even if two such contracts had a common point of juncture, unless the two contractors were required by their separate contracts to liaise with each other to ensure the two works satisfactorily joined up.

A collateral contract might deal, for example, with a contingent situation contained in the main contract. A contractor may accept the latter, provided that the Employer first made a roadway to the site. The contract with the road-maker, a third party commissioned to build the road, would be a collateral contract. Similarly, the contract direct between the Employer and a nominated subcontractor to which we drew attention in Section 1.6, would be an example of a collateral contract, allowing the Employer direct access to the subcontractor if certain situations arose. If the Employer arranged for a finance house to pay his main contractor's bills direct, the contract with the finance house would be collateral with the main contract with the contractor.

A collateral contract will not amend the terms of a main contract which is usually between different parties and often running a separate existence. The bargain it covers cannot be meddled with by one of the parties without the acceptance and participation of the other. Nevertheless, a collateral contract is a complete contract in itself, and it must have all the usual requirements to be valid; for example, a separate consideration must pass both ways between the parties subscribing to it, and this cannot be part of a consideration offered and accepted in connection with another contract. There are a number of diverse occasions on which a project manager may come across collateral contracts, but they are mostly legal ones (e.g. dealing with an agent whose agency has been unexpectedly terminated by the death of his principal); they will usually require professional legal attention and are not considered further here.

1.8 VARIATIONS TO CONTRACTS

Contract variations can refer to two situations with widely differing implications and effects for the project manager: these are variations to contract terms, and variations to a technical specification accompanying a contract. The former amends the wording or the meaning of a document forming an intrinsic part of a contract. Consequently it alters the terms of the bargain between the two parties, which is something privy to them and which they alone can change. One way of doing this is by initiating a separate Agreement so to introduce the variations, or they may determine the old contract and replace it 'as of immediate effect' with a new one which includes all the variations. No project manager or other official can make any such changes by virtue of his appointment under the original contract, even if he is named in it to take on sole management. Naturally anyone can be given special powers (or Power of Attorney) to represent one of the parties (or a

Board of Directors) at, and for the express purpose of, concluding a new contract. Anyone so doing (and each person if there are more than one) must be given a formal letter of appointment, signed by the party represented, and they must make sure their role is duly recorded in the new documents they sign. In other words, they must:

- ensure they have precise instructions from their principal as to what they must undertake and sign;
- only carry out the task for which they are appointed, and nothing else (either connected therewith or entirely separate);
- make it clear in at least two places in the documents they sign, that they are acting as the Agent (with or without Power of Attorney) for the party binding their principal;
- make it clear they undertake no rights or obligations under the contract on their own behalf;
- conclude no Deed (i.e. no document under seal) (it is unlikely, in any case, that a corporate body would allow its representative to affix its corporate seal to a document – the Articles of Association would not allow it to be delegated, even temporarily).

(For 'Power of Attorney', see Section 4.2.)

When variations to contract are extensive, it is usually wiser to rescind the existing document and to replace it entirely. A minor change can lead to associated changes in a number of closely related clauses (e.g. a change of delivery point from 'to site' to 'ex-Works' could easily make 14 changes elsewhere by the time such matters as 'possession', 'insurance', 'payment terms', 'liability for repair', 'installation', 'payment for delivery', and so on, have been brought up to date. One new contract could embrace them all).

Any amending contract must follow all the usual rules for a contract, and considerations of value must again pass both ways. Often the changes themselves are of mutual benefit, and can of themselves benefit both parties; however, the respective items must be spelt out to show clearly that both considerations are present. If they do not, then the unaffected party must itself forgo some benefit, give a new service, or provide a sum of money as *quid pro quo* for the advantages it gains by the changes. A new contract (or a contract of variation) will normally be given the same legal force as the previous one (i.e. it will be under hand or under seal like the supplanted one). This is not essential, but can otherwise lead to difficulties later on, for example, when one remains valid but the other expires.

1.8.1 Variations to specification

Most engineering contracts include a so-called 'variation clause' under which a project manager (or his equivalent in some conditions, the Engineer) is empowered to vary the specification of the works within certain limits. The changes must have a clearly defined maximum value, above which the change cannot be made without the contractor's prior acceptance, but below that figure he is obliged to accept the change provided that it does not imply a change to the contract. A change of design must never mean such a change in the contract works that it is really a new contract or one for a different purpose.

The arrangements which a project manager must come to with his Employer (who has to pay for the changes and the new works) is a separate matter, and depends to a large extent on what freedom, if any, the project manager has been given in such matters. He may be allowed to make changes to specification on his own authority up to a specified amount. Thus errors in drawings or mistakes in calculations which first come to light as work proceeds or the extent of earth-moving involved and similar matters can be readily rectified by a standing variation procedure with a minimum loss of time, while necessary changes to paper-work are set in motion. Corresponding changes to the remuneration of the contractor have to be made at the same time, and by the nature of the changes most frequently made, this can readily be done by changes to the extensions of items already in the bills of quantities. Sometimes, in addition, the schedule of rates must be extended by the project manager agreeing revised rates, the so-called Star Rates.

Variation clauses of this type are often not included in contracts for the supply and installation of plant, where responsibility for design rests with the supplier. The Employer's specification is usually functional, and the cost of even minor changes by the Employer is to be avoided on the grounds of expense. Once the supplier has started layout and ordering, any change of design may mean virtually starting again from scratch, involving not only shop drawings, but shop and machine allocations, test facilities, and the like. There is also the wasted effort 'so far' to be paid for by the Employer who, in view of the heavy expenditure involved in total, may retain himself all authority for making variations to such contract specifications. However, in working on building or constructional projects there may be many unexpected setbacks (e.g. the discovery of a poorer soil at foundation levels, or a miscalculation of the work involved in a land-levelling exercise), and it is here that the variation clause finds its greatest use.

There may be many tens (if not hundreds!) of changes of this sort to be made during site work, without any change to the basic nature of the contract works. We shall be dealing with such procedures in the third volume of the present series when we consider the control of site work on the project, but they must always be controlled by the project manager himself, who alone is responsible to his Client for the cost of the project. Changes can, however, be incorporated rapidly without sensible loss of working time; corresponding changes to all forms of records and book-work can then follow as soon as possible. They must never be forgotten, and none must be omitted from the many required by each change.

1.9 PROPERTY IN, POSSESSION OF AND OWNERSHIP

Possession, we are told, is nine tenths of the law. This is not true (or the important point beyond nine is abandoned). The law has to be precise and, equally, legal documents which involve these features must be precise as to the point at which a change occurs.

Property in a thing has a number of legal meanings depending on just what that thing is. In general, 'property' means 'a right of value can be transferred from one person to another'. Speaking casually, this means much the same as 'ownership': 'I own the thing and have the right to sell it', and this is normally true when talking of a tangible item such as covered by the Sale of Goods Act (1893) and its amendments, or engineering works.

At law, 'property' has several other meanings, many of them specialist (including land-tenure, real estate, proprietary rights, and so on), but these are usually of minor importance in the life of an engineering project manager, and can be left at that. It does not follow that the Employer's legal advisers will be uninvolved.

Tangible property (including chairs, cars, pens or butter) can form one category of what is called 'personal property'; the other category is called 'rights in action' and refers to such things as debts, patents, licences, copyrights, and the like, in which the property-owner's 'rights' are only realizable when they are enforced by an 'action' – i.e. legal action in the courts.

Possession of means what it says, nothing more; it is defined as 'having a certain degree of direct physical control over'. However, the 'state of mind' of the person concerned has also to be taken into account, it does not give the right of 'property in': I have your pen in my pocket. . . . *You own* it, even if you are far away; *I possess* it (i.e. I have no right to sell it to a third party). This, then, is the usual meaning of 'possession' when dealing with engineering contracts, but it must not always be taken at face value. I might own a piece of land, and later discover a lode of copper-ore on it, buried beneath the surface and hitherto unknown. Do I own the lode? Or do I possess it on someone else's behalf? How is my position different if a bunch of terrorists bury a cache of arms unbeknownst to me on the same piece of land? Am I in 'illegal possession' of them? Here the fact of 'being in a certain degree in direct control' of them is still basically true; I could keep the real terrorist owners out and remove them myself. So, in some circumstances, a modified version of the definition has to be used, but this is not usually the case in straightforward project work by a project manager: the Employer's legal advisers are much more likely to be brought in, should a modified definition be under consideration.

Ownership, for most engineering purposes, is synonymous with 'property in' but it is not widely used as a legal term, and it is usually given a specific definition whenever it is employed. For a casual use, it can be taken as having property in – i.e. 'I own it and, therefore, have the right to sell it.' Legally it is always better to use some recognized term, or if 'ownership' has to be used, it should be defined to say what it means in that special case.

What is obvious is that, in any contract which involves the transfer of works from a contractor to the Employer, it must be made quite specific and clear at just what point the transfer takes place. Prior to that point, the contractor/seller owns the thing, afterwards the Employer/buyer becomes the owner. He can sell it, hire it out, claim its value as his own, and of course he should insure it himself. If I buy a packet of cigarettes, the situation is fairly obvious: I pay over the money with one hand, and take the cigarettes from the shopkeeper with the other; the transfer takes place at the time of payment and delivery. But what about the case of a building put up for my eventual use? What about payment by monthly payments, or by agreed instalments? What about 'retention money' when I actually complete payment 12 months after I have taken the building over? What about delivery made by instalments?

The passing of property from the seller/contractor to the Employer is equally important. Practical matters like insurance must be affected without a break by one side or the other. When can a buyer/Employer first sell the works to somebody else, without the contractor saying, 'Hey, they're mine'? It is clear that some point at which the works, or parts of them, transfer from the contractor to the Employer must be agreed and incorporated into the contract they sign. There is no hard and fast rule, but the Employer must be certain that he is getting what he ordered, and therefore some form of completion tests must have been satisfactorily made. Equally, the contractor must be happy he has got his contract price for the things transferred before he can let them go – or at any rate, the majority of it. With items of plant or equipment, the best point is often the end of site tests following erection by the supplier on-site. If erection is not involved, often successful works tests are acceptable and the point of transfer of property is on delivery to site.

Credit terms of payment can often prove difficult, but these are usually treated as separate. Thus a seller becomes due to get payment on delivery (say) but he might be happy to give the Employer credit terms, and so get his money at intervals later: his transfer on time is not affected by this accommodating action. Whatever the arrangement made, it is important there shall be agreement at what exact point the Employer takes over, be it machinery or occupation of a building. His contract must say so, and it is enforced by law.

2 Contract legality, offers, acceptances and considerations

2.1 LEGAL REQUIREMENTS

At the beginning of Chapter 1 we briefly itemized some of the main characteristics which are shared by every contract to differentiate at the outset between a contract and a simple verbal agreement which two friends might reach with each other. In the chapters which follow we shall go more fully into these characteristics, and add others. We introduce factors in respect of them all. These are matters on which a good project manager must satisfy himself before he is able to submit for signature by his Employer any draft contract document that he has been considering. A mistake at this juncture could cost the Employer dearly in extra expenses and undue delays when such a contract comes to be executed later in the project programme. A purported contract must have the following features if it is not to be so vulnerable at a later stage of the project:

- The subject-matter of a contract must be within the law, as must be the actions of the parties to it.
- The persons concluding a contract must be legally entitled so to do.
- The parties must both intend their contract to be enforceable at law, with consequent penalties.
- There must be complete and voluntary agreement on the contents of the contract between the parties thereto.
- A consideration of value must pass both ways between the parties.
- There must be no mistakes, misrepresentations or any undue influence between the parties which might result in their contracting between one another when otherwise they would not have done.

We deal with the factors leading up to the concluding of the contract in this chapter and those affecting a signed contract in Chapter 3. All are matters which a project manager must examine. His task is of managing them later, without them 'coming apart' and causing unexpected trouble. His own effectiveness as a manager will be clearly exposed to his client later on; he must get things right from the outset.

At this stage, it might be noted that in the eyes of the law a 'person' may be either an individual or a corporate body duly represented by an authorized official, who can commit it to the obligations of a contract. Thus the term can include a commercial company and its directors and a national authority and its governors, as well as a government body and its council. Each will authorize its permanent staff to commit it as a body; some may require such commitments to be ratified subsequently in accordance with their internal standing orders. This is especially the case whenever public funds are involved.

2.2 THE LEGALITY OF CONTRACT SUBJECTS

Some contracts may be forbidden by a statute or by common law and are of themselves void. Others may be illegal by reason of certain undertakings they contain, and these have to be considered by the courts and their degree of voidability determined. It must be made clear right away that here we are dealing only with the contractual aspects, and crimes which result might be dealt with under other statutes or Acts and admit heavy punishments or penalties. It must not be thought that any decision under contract law necessarily represents the only trial which parties may have to face (see Section 3.1).

Contracts declared 'illegal' mostly offend against what is known as 'public policy'; for this reason, they are not only void, but declared 'illegal' of themselves; chiefly they are as follows:

- Contracts forbidden by statutes dealing with other matters (e.g. Restrictive Trade Practices under Part II of the 1956 Act).
- Contracts to commit crimes, etc.
- Contracts damaging to the national safety.
- Contracts impairing the nation's relations with foreign powers (e.g. inciting warlike actions).
- Contracts to compromise the operation of the nation's law measures or courts.
- Contracts leading to corruption of public officials.
- Contracts to avoid paying due rates or taxes (including payment of income tax).
- Contracts to promote sexual immorality. (No comment!)

In some cases, contracts which are not illegal in themselves may contain illegal clauses which can be disassociated from the rest of the contract (known as 'severable contracts') the objectionable part can be declared void by the courts, but the remainder confirmed as valid. Thus, for example, a contract to ship goods might include a clause which implied overloading the ship; the objective of transporting the goods might remain valid, but the 'overloading' clause declared 'void'. If the illegal clauses are not severable without seriously altering the obligations in the remainder of the contract, then the whole contract may be declared void, even though it is not illegal of itself.

From the project manager's point of view, the most important contracts are those which deal with wagering and in the restraint of trade. The latter might include restraints placed on individuals by their contracts of employment. So long as a man does not disclose trade secrets or unique methods belonging to his previous employer, an employee cannot in general be prevented from taking other work in either the same or a related field. However, 'restraint of trade' does not necessarily apply to all such re-employment, especially if the restrictions complained of are the 'accepted and normal currency of commercial or contractual relations'. The test is, as usual, one of 'reasonableness', but a company or a 'person' should not be required to undertake any contract which would itself restrict his right to trade fairly and openly. It is not likely the majority of illegal contracts will come the way of a project manager, but severable ones are not uncommon in engineering practice. Here it is usually a matter only of one or two offending clauses, and even these should not occur if the project manager is aware of their voidability and excludes them from the document before it is recommended to the Employer and signed as a contract. It is, of course, for this reason that we have included them here and spent some time considering them.

2.3 LEGAL CAPACITY OF PARTIES TO ACT

Certain classes of person are not allowed to make contracts and any they might make are voidable:

- Minors – i.e. people under 18 years of age (but see the exceptions, below); they used to be called 'infants' but since this word had a different sense in general usage, it was discontinued.
- Persons suffering the effects of drink or drugs.
- Persons suffering from mental derangement.
- Prisoners serving a sentence.

The first was designed to protect minors from their lack of knowledge of the world and its ways. The following two (drunkards and lunatics) are those sufficiently abnormal not to appreciate the full significance of the action they are taking. It is not difficult to recognize a drunkard in such a state, but it might not be so easy in the case of a person temporarily deranged, especially on short acquaintance. Naturally, if he makes a contract and his state is subsequently appreciated, it might be possible to prove this in a court of law and have his contract voided.

As we have already indicated, there are exceptions in the case of minors. They are not allowed to trade (buy and sell, for example), but they may order for their own use and contract for the supply of 'necessaries in goods or services suitable for the maintenance of the mode of life to which they have been accustomed'. 'Necessities' have never been legally defined, and it is left to a court of law to say whether something ranks as a necessity or not. Certainly, sustenance, clothing, bedding, education, medical attention, etc. are all included, but what about, for example, a man-servant who might be seen as a necessity' for an affluent minor, but surely not for a working lad? Or a supply of whisky or brandy to drink?

Another matter on which a minor can legally contract is the contract of employment, provided that it is 'on the whole beneficial to him'. Thus a project manager must watch the conditions, both physical and economical, under which he permits contractors to employ minors in running their contracts.

In spite of what general opinion might say, a parent or a guardian has no responsibility under contract law for anything a minor in his care, may legally contract: he is on his own. A contract 'to trade' is automatically outside the law and can be voided at the cost of the second party (who is expected to have known better!). A minor's parents or guardians may have responsiblities under some other law, or one they have assumed on his behalf, but that is outside our present interests. However, what is important, and is stressed, is that the other party to a contract must be aware (or could reasonably have been expected to be aware) of the status of the prescribed party at the time the contract is made: this applies mostly, of course, to drunkards and persons mentally deranged, but it could apply equally to minors.

Altogether apart from what we have said hitherto, the legal capacity of the parties acting on behalf of one making a contract has a bearing which a project manager will often face; that is, the question of whether the person with whom he is dealing is entitled to commit the second party in the way it is proposed he should. In general, it may be assumed to be so with public bodies such as councils, national authorities and the like, and every official whom one would expect to be properly authorized to commit them (for example, Chief Engineers, County Surveyors, Chief Purchasing Officers, etc.). Secretaries, and so on, would never be expected to commit their public bodies and their signatures would carry no weight. Project managers are not expected, in cases they consider valid, to carry out further investigations to prove their assumption. However, it is normal for all such bodies who spend money from the public funds to have standing orders internally for an authorized council or committee responsible for the handling of the funds to confirm a contract made by one of its appointees (see Appendix 8). It substitutes a sealed Agreement ordered by the body in session. A project manager can always expect his signed contract to be substituted shortly afterwards by a confirming Agreement; however, usually this will not alter the terms of the contract in any way, or its effective dates. It is purely a method whereby a civilian-elected body can confirm an authorization to act which they have given a permanent employee. In any cases of doubt, the claimant should produce either his letter of authorization or a more formal and legal Warrant of Attorney. We must confess that the only time we have queried the credentials of an unknown individual, he (without giving any official appointment) signed a contract worth many hundreds of thousands of pounds on behalf of his company. It turned out to be the Chairman of the Board, and a well-known peer of the realm, whom we never had occasion to meet during the course of extended negotiations. It is unlikely a project manager will suffer from a red-face so easily!

In the UK the situation regarding who is entitled to commit a limited company to specified obligations is rather a thorny one. Hitherto the powers of a company were determined by its Memoranda of Association, a legal document required of all companies registered under the Companies Act 1948. It defined the objects for which the company had been set up and prescribed the purposes for which its Board could spend the capital subscribed by its shareholders. Certain officials were entitled to authorize certain activities, with in some cases a limit on the sums of money they were entitled to commit.

Negotiators of contracts were happy to apply the same criteria as with national bodies, that is, if a person should by all normal business standards have the power to commit his company, then his agreement to a contract sufficed without further investigation. At the back of one's mind lay the fact that his company was subjected to stringent accountancy rules, and that the money for spending was firmly limited by the shareholders and not supported in some 'unspecified' way by government funds. Companies had to risk their own money, not the taxpayers'. Such an arrangement worked very well, and there are virtually no occasions when, in such circumstances, a company did not support the signature of a senior employee.

The situation has recently changed, partly in favour of the would-be negotiator and partly the other way. Since 1972 the UK has come into line with its fellow-members of the European Community and now accepts that 'any transaction decided on by its Board of Directors in good faith shall be deemed to be one which the company is permitted to enter into'. In other words, the company is not now restricted in its activities by its Memoranda of Association, but it is allowed to take on anything which its directors (and through them its shareholders) may decide. This would tend to widen the sphere of activities of a company and allow them to engage in some 'experimental' side-shows (always remembering that they are at risk if their experiments should fail). This might lead Employers to feel they could be experimented on by their contractors when they carry out a project. On the other hand, we have the fact that recently there has been a big increase in the application of Quality Assurance, where an Employer can require proof from a contractor that his certificate of good 'experience' is to be trusted; in other words, he obtains an *assurance* that he will get a good result from his contract. This works the other way, restoring the balance and reliance on the project manager.

In the case of contracts overseas, it has always been generally accepted that foreign parties include in each contract (often in two places – in the opening recital and also at the point where the companies actually sign at the end) the names and appointments of the persons whom they are authorizing to sign on their behalf. Such a procedure removes any doubt as to whose signature they are prepared to accept as binding on the organization, always assuming the individual concerned remembers to produce his personal identification. We have never known this to be asked for, or given!

2.4 INTENTION FOR CONTRACT TO BE SUBJECT TO LAW

Here the parties to a contract shall both intend that their obligations and benefits shall be supported by being subject to the law in the UK. In those affairs which might be described as social or domestic (and which are not likely to affect engineering work), the terms in the conditions of contract usually imply such an intention (e.g. quoting the law to apply, references to legally awarded damages, etc.), but if these are not present or if there is any cause for doubt in the matter, a clause clearly stating the intention should be included. This expressly states that it is the agreed intention that all contracts should, if possible, be supported by the appropriate law.

Thus, in the commercial world, it is always taken for granted that a contract is to be legally enforceable, and the courts always accept this without requiring either party to produce any further evidence. The legal processes and formal judgments can consequently always be properly followed. A defendant can, nevertheless, plead that his contract is *not* intended to be supported by the law if he can produce evidence to this effect by the wording used in the contract itself, or in any other document. A contract does not have to be legally enforceable if it contains an agreed declaration to the contrary. The courts will support such an *agreed* statement.

Collective bargaining by trade unions on behalf of their members is nowadays intended (both by the unions and by the employers of their members) to be binding at law on each member individually or collectively. It is a condition of their membership and of their employment. However, in the Industrial Relations Act 1971 it was agreed that any agreement made under the Act between the parties must be recorded in writing. It all comes back, as usual, to a matter of proving just what was said and agreed at the time!

To sum up, if and when *both* parties agree and write into the contract that it shall not be regarded as subject to law, then the courts will accept this record and uphold the decision of the parties.

2.5 MUTUAL AGREEMENT: OFFER AND ACCEPTANCE

2.5.1 The offer

We have already explained how mutual agreement between the parties is most readily demonstrated to all concerned by an *offer* made by one of the parties, which is *accepted* by the other without conditions or modifications. The offer must naturally embody the whole of the terms of the agreement which are to appear in the final contract. Frequently the process starts one step earlier (especially when competitive tenders are invited) by the Employer issuing in the form of an enquiry all that he expects to find in any contract that he eventually signs. The appointed project manager will agree the terms of the enquiry (indeed he may often be the composer thereof) and it will embrace the standard and special conditions of contract, the detailed specification, the Employer's regulations for his site and the exact form of the offer to be made. All tenders received are then in the same form, embrace the same details and are strictly comparable. Each competitor will, of course, enter his price and any special features he is not able to meet, but little more. The system does not work so well if the contractor has his own ideas as to how the job should be done and merely alters the papers he receives accordingly. The Employer then has to examine his offer word by word to determine just what features the contractor has failed to follow. There is no other way of finding out what effect they might have on the price quoted. It might produce a low price and the manager appreciates too late that an essential feature has not been included. To make his task less onerous – and still maintain some comparability between tenders – his enquiry should either:

(a) insist all tenderers bid for the actual work as set out in the enquiry, before they add their own second tender price, explaining how their scheme adds to (or falls short of) the Employer's; this solution breaks down, of course, if the contractor is simply not organized to follow the Employer's scheme (he may not have the requisite machinery, for example), and consequently he is prevented from bidding a meaningful price; or

(b) include a pro-forma with the enquiry on which the contractor lists all deviations from the Employer's scheme and includes the relative merits of his substituted proposals. The project manager may then take advantage of a more sophisticated approach, or make a financial adjustment to his set scheme – at least he is spared the need for a detailed examination of the competitor's offer; if he does not care for the tenderer's outlook, he quickly removes him from the competition.

Legally the Employer is not obliged to accept any of the offers he receives, still less the lowest one, even if he approaches a single contractor. He may wish to negotiate with several offers and, in the process, produce a number of counter-offers. Eventually it may be rather unclear just which party has offered what! In such circumstances, it is frequently better to invite the contractor to start again from scratch by submitting a new offer 'so far'. Eventually an agreed offer and acceptance results, and a contract is formed. Competitive tendering is usually straightforward; it is relatively easy to recognize a genuine offer, and what form, the acceptance has taken. What is not quite so easy to identify in detail is just what the offerer has proposed. Is the resulting contract a valid one, or is it not? Is the offer definite, or does it leave too much to unspecified generalities or (worse still) to uncertainty? Unless the offer is in reply to an enquiry from the Employer (which the project manager will have vetted before dispatch), such points may be obscure but must always be considered before the project manager makes any decision or takes any action. In competitive tendering each competitor must be tendering a price for a common specification of the contract works, so that direct comparisons can be made.

2.5.2 Invitations to treat

Apparent offers, made to the public in general, should always be regarded merely as 'pseudo-offers', the contractor proposing that he and some second party, unnamed, should get together and try to arrive at an agreement and thus a contract. He himself is not making at the outset an offer to contract, open to acceptance by anyone interested. It is a general 'invitation to treat'.

Thus, at law, a display of goods in a shop window (even if the price is attached to them) is not an open invitation to accept the offer by some passer-by and form a buy–sell contract. It is an 'invitation to treat': the shopkeeper may (if he wishes) accept the offer of someone interested in making a purchase, by agreeing to negotiate and make a sale. The same is true of an advertiser who proffers machinery in a trade journal at a stated price and delivery: he is inviting the public to negotiate with him and come to an agreement. It is the applicant who makes the offer, and the advertiser who agrees to deal and makes the sale. Neither he nor the shopkeeper is obliged to deal with any customer who presents himself, nor to settle for the price he has signalled in his invitation to treat.

Often such a situation is difficult to distinguish from a true offer, capable of being accepted and forming a contract. The intent must be thought through carefully; perhaps the safest guide is whether the so-called offer has been made to an individual person or firm, or whether it is in the form of a general broadcast to the world at large. This may not be a very sound legal approach, but as a first step it is the easiest one for a project manager to make.

There are some extensions to the principle which we might look at with advantage. Take a supermarket where priced goods are displayed on shelves for customers to pick. Certainly, a contract between the customer and the owner comes into effect at some point, but just when and how? In a test case it was ruled by the court that the point-of-sale was the cash-desk. Here the customer 'offers' to buy the goods he has in his basket, and the owner (through his agent the cashier) accepts the offer and agrees to sell. The owner has no obligation to do so, or to sell any article at any preconceived price. But to save negotiation in the queue, the shelf-price is normally maintained, and a sale is rarely refused. Similarly, at an auction sale the client makes an offer by rubbing his nose (or some recognized signal) and the auctioneer accepts it by banging his gavel on the desk, neither needs to speak a word. The notices of the auction and the display of goods are 'invitations to treat', and all lower bids (offers) are refused by the auctioneer when he accepts a higher bid. Just as in competitive tendering but in reverse, highest wins!

2.5.3 Legal termination of an offer

An offer made with the intention of forming a contract can cease to be effective in any one of a number of ways:

(a) *By lapse of time.* Most offers include a specified term of validity, i.e. the offer remains open for acceptance or rejection within a specified period. It then lapses purely by efflux of time unless it is renewed in writing by the offerer; in effect, he is making a new offer in which he may amend his price, or its terms, or even he can refuse altogether (if, for example, some necessary facilities at his factory have meantime ceased to be available). If the original offer contains no stated validity period, it may only be accepted for a 'reasonable' time. A wise project manager would first confirm that it was still alive for acceptance. There is no definition of 'reasonable', of course, and it must depend largely on the circumstances: the offer might concern perishable goods, or something which is only available 'on the market' for a very limited time. In such a case, a 'reasonable time' might be something very short – a matter of hours or days. Or alternatively, in an inflationary climate, the mere rise in his costs may make it necessary for a contractor to withdraw his offer and re-issue it with an increased price. This could be the case after even short validity periods.

(b) *By refusal.* The recipient of an offer cannot go back on his decision once he has notified refusal of the offer to the offerer. He cannot have a change of mind. Once an offer is refused, it becomes non-existent. The making of a counter-offer by the recipient might legally form a refusal, and the offer concerned could be considered void. However, this situation is not entirely clear as any attempt to negotiate could be regarded as making the offer non-existent. In practice, an attempt to negotiate on some detail contained in the original offer is not regarded as a refusal (indeed it implies continued interest by the recipient). A counter-offer on completely different lines would signal a refusal of the offer, replacing it by the new proposition.

(c) *By revocation.* Under English law, an offer can be withdrawn by the offerer at any time prior to its acceptance. Revocation must be communicated to the recipient, and it does not come into force until the latter has actually received such notice of withdrawal. Thus a notice sent by post would not become effective until it has actually been delivered. The 'communicator' does not have to be the actual offerer; *he* could have fled the country, and left it to a friend to clear up the mess!

(d) *By non-fulfilment of a condition.* If an offer has been made subject to some circumstances becoming fulfilled, it cannot be accepted until this has happened. If it looks like not happening at all, an offer can become void by common consent; it is not unusual therefore for such an offer to specify the date by which the requirement must be achieved. As examples we might quote (i) any goods concerned must stay in their original condition – i.e not become rusty; (ii) the recipient might be expecting an inheritance with which to finance the contract; and (iii) the well-known phrase 'subject to contract' might be used. This last example leaves the writer without liability, should the contract never be made. Intending house-purchasers will recognize the phrase. Although it is not likely to affect project managers who have an engineering contract to control, it might be of interest merely to show how frequently one comes across exceptions to any rules laid down elsewhere. There is one class of advertisement which has been ruled as not an 'invitation to treat' (as discussed above); but a true offer of an agreement 'subject to condition', concerning advertisements offering rewards for the recovery of lost or strayed goods, or for 'information leading to the recovery of stolen property and the apprehension of the thieves'. The condition is, of course, the fulfilment of the recovery or the information by the claimant, who is then inferred to have fulfilled it and accepted the advertiser's offer. A contract then exists and the claimant has a legal right to his promised reward.

We have drawn attention above (item b) to the danger in some cases of an initial attempt at *negotiation* being misread as a *refusal* of the offer. This is very real in some circumstances, without any suggestion by the Employer that he wants the offer cancelled. Usually the two parties get together as soon as the Employer's enquiry is issued, so that any more sophisticated schemes can be discussed and negotiated before the legal offer is composed and sent. Its contents will then be largely in accordance with the Employer's wishes, and major changes will no longer be needed. The risk of premature cancellation is then considerably reduced. Nowadays the same sort of pre-discussion is met with in some types of two-stage tendering, in which a preliminary enquiry is broadcast and a subsequent enquiry in accordance with the second party's wishes is then issued by the Employer. Excessive post-negotiation is avoided.

2.5.4 Acceptance

A contract is formed only when there is complete mutual agreement between the parties. This is often demonstrated (not only to the parties themselves) when one of them (usually the Contractor) makes an offer to the other (the Employer) which is unconditionally accepted. Any acceptance including a 'but' or an 'if' or different arrangements for paying, or a different scope to a specification, or any similar new feature, is automatically not an unconditional acceptance. At law, it is another offer, a counter-offer which, in turn, needs the agreement of the other party. Sometimes, after receipt of an offer is arranged, there will be extensive negotiations between the parties on matters which are unacceptable to the receiver. In the end, the agreed document bears no relation to the original offer. It will be found advantageous to both parties to 'start again' with a new offer, embodying all the freshly negotiated details, and which can be unconditionally accepted.

An 'offer' may be originated by the offerer, or may (as is often the case with engineering contracts) be 'suggested' by the enquiry sent out by the Employer or his project manager. This includes all the contract conditions which are to apply, in fact everything except the contractor's price, delivery and omissions; these are added, and the whole document then comes back as the offer. If the contractor disagrees with any of the suggestions, he must say so in his offer, and the Employer can then accept the changes or negotiate an agreed alternative. An acceptance must always fit the offer, exactly.

This requirement of the mutuality of such agreement between the parties is not unlike the background logic of the old farmer who bequeathed his farm 'to be equally divided between his two sons, A and B'; they could not agree what was a fair division. The local judge soon solved this by appointing A as the elder to make the division, but B as the younger was then to have first choice! An offerer will not make a biased offer when he knows the Employer can reject it!

Once it has been established that an offer is a true one, and not merely an invitation to treat, an acceptance is not difficult to reach. Much more troublesome is the need for this acceptance to be communicated to the offerer, as indeed it must. There is no validity in what the other party 'has in mind' or 'intends to do'. He must make up his mind and tell the first party. A contract is based on mutual agreement; how can there be any contract if such agreement has not been communicated to the one making the offer? It is almost certain that such communication must be made by the person to whom the offer was made, or at least by his appointed agent, such as the project manager, if he has by that time been so appointed. Thus acceptance, in the general case, only takes place when the offerer is actually told of the acceptance, face to face, by telex, by a clear telephone line, or similar means. There is an anomaly when the post is used, and this has not been satisfactorily settled to both parties. The courts have held that, in the UK, an acceptance is valid as soon as it is committed to the post, even though the offerer may not hear about it for several days. The Employer, for example, has made up his mind, has committed his decision to writing, has posted it (so he cannot alter it) and has therefore committed himself and a contract has been formed. Such contract is already in existence, and he can be sued for breach of it, should he in the meantime change his mind. Maybe he has received a better offer? So he telephones the offerer (and thus overtakes his letter) saying he rejects the first offer.

The courts then say, 'Sorry, old man, but you already have yourself a contract. Break it at your peril.' The situation is not nearly so happy if the offerer withdraws his offer, and his revocation crosses the acceptance in the post. The situation has never been completely settled in English Law, but is probably a matter of intention. If the offer is such that it must be answered 'Yes' or 'No' within a day or two, a reply by post might be regarded as too dilatory, and the revocation is allowed to stand. If offered by telex, a reply by telex might be expected; if offered by second-class mail, the other party cannot be expected to rush around madly, and the general rule as regarding posting his decision would stand. Such rule holds, even if the letter is delayed in the post, or even if it is completely lost!

Acceptance can be implied by the actions of the second party, even without a formal acceptance. Thus an offer to supply goods at such-and-such a price can be considered 'accepted' if the recipient receives the goods and takes them into his store. Delivery in kind can rank as acceptance. The auctioneer's gavel (which we have referred to elsewhere) is acceptance by act, not word. Even acceptance by action must, however, be communicated by some written statement, *post facto*.

The following summarizes what we have already said and acts as a reminder on the subject of acceptances:

- An acceptance must fit the offer exactly. There must be no new suggestions or amendments.

- An acceptance must be communicated to the offerer.
- If done by telephone, a witness should be present who can later testify as to time and date.
- If communicated by telex, the telex print-out acts as its own evidence of receipt.
- A contract dates from the time when the offerer is told of the acceptance, except in the case of posting, when it is the time of handing over to the Post Office or the post-box, NOT to a postman delivering letters!
- If sent by post, a certificate of posting is important.
- In some circumstances, acceptance is given only by action, but this still must be communicated to the other party and preferably confirmed.
- Silence does not imply agreement or acccptance.

2.6 CONSIDERATION

We have already had occasion to refer to the factor of 'consideration', which must flow both ways between the parties to a contract and bring to it its essential feature of a bargain. A consideration can take any form (money, advantages, service, etc.) and its value is immaterial provided that the party receiving it has agreed that it is adequate in the circumstances; thus, in engineering contracts, it is frequently money for work done or for goods supplied. The essential of a bargain is sustained: 'You give me something, and I will give you something in return, the value of which you consider adequate.'

A wealthy grandfather might engage himself to give his grandchild a sum of money in later years. This is a promise, it is an obligation undertaken, it might be enforceable at law, and it certainly has something of value passing one way. But it is *not* a contract: no consideration passes between the child and the grandfather. The legal 'bond' is a promise and made under seal. It is a common mistake to assume that a contract made under seal requires no consideration to pass in both directions. All contracts (both under hand or seal) require the aspect of a bargain, with something of worth going both ways between the parties.

The 'consideration' need not be one of money: it can be a promise (i.e. to pass something of value in the future) or an undertaking to pass something of value immediately. It can rarely be some value that has already passed in the past (i.e. 'for favours received'). An exception to this dictum is where a past favour was made conditional on the present contract. Instead of cash, a consideration can consist of 'some right, interest, profit or benefit accruing to one party, or some detriment, forbearance, loss, or responsibility given, suffered or undertaken by the other'. In short, a benefit one way or a detriment the other. Reliance on any past consideration always has the risk that it might have arisen from a breach of the law or of another contract, at the time; this can no longer be checked, and (in any case) a breach of a contract by one party is not an excuse for any breach by the other party! If one party (say, a contractor) fails to carry out his undertaking to deliver certain goods, the other party does not have to pay for them 'on delivery', but he must be able and willing to pay at the time they are due to be delivered. Otherwise he too will be in breach of contract with all that that entails.

From the point of view of the law, a consideration can be quite trivial. The courts have no interest in whether the two considerations are of equal (or similar) value. There is no question of a balanced *quid pro quo*. The essentials are that both considerations are recognizable, specific or factual, and acceptable as such to the receiving party. What is a 'good bargain' and what is a 'bad bargain' does not concern the law; the two parties must stand by the bargain they have made. A number of points require to be checked before the project manager can put up a contract to his Employer for signature:

- A consideration must be clear and concise; phrases like '20% of any profit I might make' do not form a consideration.
- Considerations must pass between parties to the contract, not between any third party, but see also the case of assignees.
- A consideration must form part of the contract itself. It cannot be a dissociated gift or a bribe.
- Generally speaking, a consideration must not result from any duty the party already owes to the public or to the other party, but there are many points of legal doubt on this issue.

2.7 LETTER OF INTENT

The title Letter of Intent is the cause of considerable confusion when, as often happens, the actual wording used is such as might constitute an acceptance of an earlier offer or, at any rate, an Instruction to Proceed. Indeed in a number of overseas countries the title is used with just such an aim and intention. The actual wording deserves to be examined closely on every occasion that the title is met with, to see whether it is actually much more than is implied by the name at first sight. It may legally be another document altogether. 'A rose by any other name, would smell as sweet.'

A true Letter of Intent should contain useful information for the tenderer, but should not be used as an unconditional acceptance (either of a whole tender-offer or a specified part). By its very name (i.e. we *intend* to do such-and-such), it cannot be a firm decision and therefore has no legal or commercial value at all.

Its main use is as an early warning at executive level, that a formal contract is on its way which may be subject to a small administrative delay in reaching them, possibly because the Board of Directors, who must confirm the contract, are not available for a few days. The executive manager can then use this few days' grace to make early appointments, select site teams, earmark plant and equipment, and reorganise his internal programmes. Naturally a letter of intent bears no authority for expenditure, and all costs are at his own liability until they can eventually be charged to the contract itself when it arrives.

2.8 INSTRUCTION TO PROCEED

The object of an instruction to proceed is to temporarily restrict the scope of a contractor's activities on a contract. He must first have his contract, and the opening of the instruction (or better still, a separate acceptance document) must establish the unconditional acceptance of his whole tender-offer. This not only binds him to the Employer for the full works, but avoids difficulties which might arise by a refusal to take them on, or expiry of the validity of his offer. His immediate activity may then be restricted, for example:

- To a temporary financial limit (e.g. while the Employer implements his financial arrangements).
- To specified parts of the works (e.g. other contractors running late, site access not ready).
- To avoid special contract conditions still under negotiation (e.g. with a third party).

Such restriction may cost the Employer money and be the subject of a contractor's claim for enforced delay or for interference with his free access to the work. An early intimation of the duration of the restriction is most helpful to a contractor trying to adjust his forces to the interim regime. Like any Employer's instruction, this one needs to be acknowledged in writing, as does the later instruction removing the restriction and re-establishing unhindered working of the contract programme. In the extreme case, for example, a contractor may vacate the site entirely without claiming frustration of contract, or an intention to revoke it; it all depends on the nature and duration of the restrictions imposed on him (see Appendix 13).

3 *Uncertainty, undue influence, misrepresentation, mistakes, frustration, etc.*

3.1 DIVISIONS OF UK LAW: THE RIGHT CHARGE

Up to now, we have been dealing primarily with those matters which arise during the formation and negotiation of a contract. In this chapter we shall deal with those matters which occur after the parties have agreed the terms of their contract and, to begin with, are both happy with it, valid or not. As other matters come along, the rot sets in, and things are not so rosy. By this time, each knows just what their contract apparently contains, and what they expect to get out of it. Nevertheless, they frequently become less satisfied (especially if things do not go exactly as they would wish): is the other party really doing his best? Are there misstatements or other matters of doubt? If only one can pinpoint the real causes of dissent, the remedy is to appeal to the law and let the judiciary decide what needs to be done. Indeed, whether the contract is of any value at all, is it valid, or should parts of it be rescinded or replaced by something else?

There is glib talk about 'sueing somebody', but what exactly is meant by this? Actually it means that one party charges the other before a court with a clear breach of some duty owed to them by the other. The court requires the plaintiff to state the particular law which establishes the duty, to produce evidence that this duty is indeed owed to him and, finally, that this particular rule of the law has been broken by the other party who is being charged. It is no short and simple matter; the court then confirms that the law invoked is applicable, examines the evidence and finally rules on what it finds.

English law is split up into several different parts, all of which affect our daily existence. Just how far a project manager can follow this so-called division is open to question, and much will depend on his legal advisers. For our immediate purposes, it is sufficient to recognize that, if a party is to be taken successfully before the courts, he must be charged with the precise breach of duty which his actions provoke, and no other. Many of the so-called divisions are unlikely to come the way of the manager in charge of a project, even if he is in some remote territory (for example, Divorce Law, Hire-Purchase Law or even Maritime Law), but there are three divisions which do directly affect what we shall be considering and which all project management ought to recognize:

- *Contract law.* This contains all the rules and regulations relating to the formation of contracts themselves and the parties agreeing them. It lays down precisely what a contract requires and how it says it. The plaintiff must quote the actual clause of his contract which makes the second party liable towards him, what he requires to be done, and what evidence he has that such liability was not carried out. He claims such awards as contract law prescribes in such circumstances. A charge under this division of the law relates solely to a contract: no contract, no case for trial; no breach, no recompense.
- *Common (or civil) law.* This deals with purely individual offences against the established civil code. There is no need for a contract of any sort; it concerns the duties which a person owes to his neighbours, and deals with all personal offences such as trespass, assault, owing money, property damage, stealing a person's goods, domestic affairs, and so on. Such an offence against common law is a 'tort'. A person having a contract can still charge a man with a tort, but the contract plays no part in the court proceedings. Similarly, the damages, etc. he may be awarded are calculated on tortious principles and not on contract statutes.
- *Criminal law.* This deals with offences committed against the public at large, not affecting any one particular individual. The charge is laid against the offender by the police or other public security organization when given appropriate powers (e.g. under military law). It deals with such things as motoring offences, widespread fraud, conspiracies of all sorts to commit illegal acts, rioting, any offences against public order (and, of course, when an individual concerned is no longer able to pursue his own case by having been murdered!). No contract is involved, except perhaps an illegal one.

Even when it is clear that an offence has been committed, one cannot charge a person with just anything; charges like 'he didn't turn up as arranged' or 'he nicked my bike' count for exactly nothing. It must be decided precisely what Act has been broken, and the offender is charged with breach of that Act and no other. Sometimes a decision has to be made between two possible laws, one (say) of contract and the other of common law: the latter may yield damages but leave the injured person with a faulty contract. There is no offence of 'fraud' under contract law! Common law is the sole option. It might sometimes be necessary to bring charges under two divisions of the law, one dealing with the offence under common law (i.e. the tort), and the other to rectify or rescind the faulty contract in a way acceptable to the courts.

3.2 UNCERTAINTY

This is perhaps the most common of the faults which lead to unhappy relations between the parties to a contract. What exactly did the parties mean when they signed such a vague contract? A contract must always mean something, and the *same* thing, to anyone who has cause to read it, not merely the two parties themselves with their background knowledge – but certainly to a court of law. They must say with certainty what it is all about, what its aims are and precisely what responsibilities are laid upon each party.

A court can rule that the whole or part of a contract is void on grounds of uncertainty after they have considered all the claims and evidence presented to them and made every effort to discover what the parties really intended at the time of signing. It will take into account any local abbreviations, common usages, trade jargon and any previous completed contracts between the same parties. It is only in this way that one can ascribe meanings to some vague phrases, 'the usual terms', 'a normal specification', 'acceptable quality', and so on. Some time ago the Law Lords established the principle that, whenever possible, the courts would seek to support rather than defeat any reasonable bargain the parties may have intended: 'We do not want to incur the reproach of being a destroyer of bargains.' Courts will therefore provide implied meanings as they think fit or may declare certain clauses of the contract void if the validity of the remainder is undisturbed.

Courts must always bear in mind, however, that one of the parties is unhappy with his bargain and is maybe seeking an excuse to get free of it. They must also consider the position of any third parties who might be affected and decide what seems best in the circumstances. They cannot 'invite' every aggrieved person to query his contract on the slim grounds that he himself does not fully understand it (the easy way out!).

It is not always necessary to treat a contract as a whole document. It may only be a matter of one or two clauses which have given rise to conflicting interpretations, and which can be declared void through uncertainty. The aim, as always, is to find out what the parties themselves really intended when they first drew up and signed the document. Voiding a part of a contract will only be done if the remainder of the parties' agreement is virtually unaffected.

3.3 UNDUE INFLUENCE

Contracts must always be based on complete and voluntary agreement between the parties. A party placed by fortune in a dominant position must not urge the other party to agree against his better judgement. The original common law term was 'duress' but this was restricted at law to actual or threatened violence against the individual at risk. It did not apply to his relatives or associates, his land or goods, nor to a corporate body like a company or firm. To a project manager it had a very limited application: even in these days of frequent 'mugging', one is seldom attacked in this way to induce one to enter a contract!

The scope was therefore extended to form the doctrine of 'undue influence' and give a much wider interpretation of the form of coercion used. It now embraces actions at law (or threats thereof), prosecutions for real or imaginary misdeeds, proposed threats or penalties for non-compliance. It is sometimes applied to an excessive use of powers invested in 'dominant' parties such as solicitors and their clients, doctors and their patients, teachers and their pupils, etc.

It is never applied to married couples (who is to say which is 'dominant' or which of the partners influences which?). Undue influence therefore is (in contract law) any threat or action which induces a party to enter an agreement against his better judgement – i.e. the agreement is no longer a voluntary one. In normal cases, the onus of proof lies with the complainant who has to show that:

- he was persuaded to form a contract he would not otherwise have made;
- influence was brought to bear on him so to do;
- the influence used was, in the circumstances, unduly strong;
- that the defendant used the influence complained of but had no statutory powers so to do.

However, if on the other hand, the person complained of had been given some legal powers under another statute, which tended to make him dominant, such as a reform-school master or the partners mentioned above, for example, he can rebut any charges by showing that what he did was within the powers granted to him and was therefore not undue; they were appropriate to the duty imposed on him and were not excessive in such circumstances.

A contract involved in a complaint of undue influence is not inherently void: if the plaint succeeds, however, it can be voided by the courts. They rule that such a contract never existed and any moneys which had already changed hands under the terms of the contract must be returned to the payer, less any such money as might represent in their view a useful asset to the plaintiff. This is one of the cases in which besides its effect on the contractual situation, there might well be offences against common law which have to be dealt with separately, and incur associated penalties. The contract-law case is very necessary, as without the due process of the law, it might be difficult to persuade the guilty party to disgorge any money he has received. In addition, the findings of the first case might provide unassailable evidence for the second.

3.4 MISREPRESENTATION

Misrepresentation occurs when a person is influenced in his decision to enter a contract by an untrue statement made to him: such a statement must:

- influence him to enter a contract;
- be one of fact: opinions, estimates, approximations, and the like, are inadmissible;
- be made by one who is able to appreciate the importance of his remarks – i.e. not drunkards, mentally deranged persons, infants, etc.

Note that the person must be influenced by the statement itself and not by any investigator's report he may get. A great many statements are made during the negotiations leading up to a contract. Some may be purely 'cock-shies' or suggestions which are dropped after a short discussion. Others may be seriously considered, even to the extent of getting into the draft of a proposed agreement, only to be omitted from the final contract itself, or any collateral contract. Some will find their way into a completed contract as one of its terms.

We deal first with the last-named, namely that case where the statement of fact becomes a term in a contract. It is then a purely contract matter. If later proved to be an untrue statement, the term containing it is, *a priori*, in breach. The party who suffers then has a right of action (as with any breach of contract) and what he may claim will depend on the seriousness of the untruth. It might affect a 'condition' or a 'warranty' (see Section 1.5). The other party is in breach, and it is immaterial the degree to which he might be innocent, or that he derived the statement of fact from (as he thought) unimpeachable sources. As far as a project manager is concerned, he is dealing with a plain case of breach of contract but must produce unassailable evidence that the statement of fact is indeed untrue.

But what of all the other statements of fact which do not finish up as a term of a contract; they can still influence a party sufficiently to persuade him to enter a contract against his better judgement (unaffected by the untrue facts). Since the statements do not appear in the actual contracts, any attack on the lines previously suggested is useless from the start; any case brought on such a basis is readily answered by a question, 'Tell me, which clause of the contract has been broken?' Yet the injured party still merits some retribution: he has suffered a setback to his project and his contract cannot save him. He has no option but to attack the wrongdoer under tort, as the perpetrator of a fraud, and claim damages on this basis. Remember the statements must comprise facts, either statements or specifications purporting to give exact figures. If they are merely thoughts or opinions, any fault which might exist lies with the plaintiff himself for paying too much importance to them.

Common (or civil) law recognizes three distinct types of misstatement, which (if they influence a person to sign a contract) are referred to as misrepresentations:

- Fraudulent statements which are made with *intent to deceive* the hearer and influence his judgement.
- Negligent statements which are made as facts without any check as to their accuracy or without caring if they are true or not.
- Innocent statements made as facts and fully believed by the maker, both from his source or from his own confirmation. They are subsequently shown to both parties to have been false.

Although the speaker of any such statement may have no ill-intentions at the time of making a misrepresentation, this is of no consequence if the hearer was affected by his statement, and as a result, entered into the false contract. The word 'negligent' in the second type has rather a different meaning from its usual legal one. Here it does not mean the non-fulfilment of a duty owed to the injured party, but has the more common meaning that the party did not take the trouble to confirm the truth or otherwise of a statement he accepted as being correct.

The whole subject may become clearer if we consider an example from an engineering project. A project manager is concerned to obtain locally 300 tonnes of hardcore. He negotiates with a seller who has a dump of suitable material in a field behind the church, which is reputed to contain at least the 300 tonnes he requires, and he has a preference for a contract which would deliver it to him. The seller knows from his weigh-bills and similar documentation that there are only 200 tonnes in his dump, but he is anxious to get rid of it, so he assures the purchaser that it contains at least 300 tonnes. If the purchaser enters a contract to obtain it, the seller is guilty of fraudulent misrepresentation – his statement is untrue and is made with intent. He hopes to get away with it!

Suppose, however, the seller has no idea how much hardcore he has and makes no effort to find out whether he has the 300 tonnes the purchaser wants, but sells the dump contents to him on the off-chance there will be enough. He makes a shot-in-the-dark and assures the purchaser there is ample to hand. He is then guilty of negligent misrepresentation, in that he neglected to try to discover whether he could meet a main condition of the contract.

For our third situation, suppose the seller holds weigh-bills from a haulage contractor who brought the hardcore to the dump originally. These show quite conclusively that 300 tonnes were duly delivered. He checks the totals and is justified in assuring the purchaser that he has the requisite 300 tonnes, and the supply contract is signed. However, unbeknownst to him, the haulage contractor has 'fiddled' the weigh-bills to support his overcharging for 300 tonnes, whereas he actually delivered only 200 tonnes. The dump certainly appears to be of 300 tonnes and both parties to the resulting contract are innocently happy. The villain of the piece is, of course, the haulage contractor and his false haulage contract. The seller has made an 'innocent misrepresentation' to the purchaser, but he may still sue his haulage contractor.

We can use the same example to show how hard it is to give real figures for the award the purchaser can get for being lured into a false contract. Suppose his contract specified that 300 tonnes were to be delivered: the purchaser is satisfied, as nothing is said about the dump in the local field, and he gets the 300 tonnes he wanted. His damages will be minimal. If, however, the contract specifies 'the dump behind the church', the purchaser will find himself short of his intended total; and besides holding up his work, this may force him to find a further supply and enter another contract with somebody else. His damages will be much greater. Note that if the offending statement was introduced into the contract, the purchaser could sue for breach of contract, and the seller would be considered in breach regardless of his intentions. He must not contract for something he is not sure of.

As to redress, the party injured can, at common law, get appropriate damages from the other party on the grounds of their shortcomings, but in the case of innocent misrepresentation neither party is blameworthy and so damages are not admissible. At the same time, the aggrieved party has a faulty contract on his hands, and this has to be put right.

With all three types of misrepresentation, the injured party has a right to rescind the contract. This he can do by simply informing the other party that he is 'avoiding' the contract, but if he has difficulties, he may equally appeal to the courts. They will order recission and also a financial adjustment to put both parties back into their original positions as though the contract had never been made. Note this money payment is not damages, but a restitution (known legally as 'an indemnity'). Note that in the case of innocent misrepresentation the court may at its own discretion (but not at anybody else's) replace recission of a contract by an award of damages. This is not a right the purchaser can enforce, but a discretion of the court. Its main purpose seems to be to avoid the not inconsiderable upset which might result (especially to third parties) when a very insignificant gain might follow from recission: a small monetary adjustment suffices in lieu.

Instead of having the contract rescinded, an injured party can affirm his contract and let it proceed as though nothing had happened. He might well do this if the net effect is small and the unhindered progress of the project is his prime consideration. He could still, if necessary, sue the offender under common law and claim his damages. Or he could just do nothing!

As is usual in English law, damages claimed at common law must be restricted to what has been actually incurred as a direct result of the matter complained of. Penal damages (i.e. those claimed as a punishment) are tantamount to a fine, and are a matter for the courts themselves, not for the plaintiff.

3.5 MISTAKES

In spite of all the checking and care that goes into the drawing up and signature of contracts, mistakes still occur. It is frequently said that a party to a contract can appeal to the courts to have *any mistake* removed. This is not true, or only partly true, since a 'mistake' in the legal sense is quite different from what one commonly associates with the word 'mistake' in everyday speech. A legal 'mistake' is limited to one which might occur in the legalities of drawing up a contract. It has always been made very clear that the law is not concerned with the freedom of the individual to agree to any bargain he chooses. It matters not if he has made a foolish bargain (a mistake perhaps?), involving him in considerable loss on the contract, an error of judgement from which he now would like to extricate himself or anything of that sort. He has made his bed, and must lie on it. The law says, 'these might be mistakes in your language, but they are not so included in ours'. It is often found during an appeal to the courts (or examinations of documents in preparation therefor) that a 'mistake' is in fact a breach of contract, or of misrepresentation or some similar offence.

Mistakes can occur in making offers and acceptances. They can be made to or by the wrong person, so that the identity of the second party to a contract is in doubt. This is no longer unusual, with present-day proliferation of 'take-over bids', 'joint developments and research' or similar co-operative schemes. It can no longer be unusual for a project manager to recommend a contract to his Employer, only to discover later that the firm in question has been taken over by his competitor. The latter might then have a stranglehold on the Employer's project and delay progress indefinitely unless a suitable 'ransom' is paid!

Similarly, an acceptance might not exactly match an offer. Both parties think they have a good contract, but legally it does not exist at all. The classic example was the purchase of a derelict ship on a specified deserted island. It eventually transpired that there was no such ship and no island of that name in the area.

If there can still be any doubt that such mistakes readily occur, the following two examples – known to the authors – can be quoted. The first relates to a large civil engineering contract at two sites. Each site had a secretary with the same surname, but one was redundant as work was almost completed. So Personnel sacked the other (busy) one, who enquired what she had done wrong to merit such instant dismissal. The second relates to the days of national fervour preceding the Second World War. Two unmarried twin boys enlisted together into the same arm of the service. They had the same next-of-kin and the same address. Their 7-figure service numbers were only one apart. Each had but one 'given' name, and in both cases it began with 'J'. The mix-up of their affairs at Records and the Paymaster was made worse when the typist so frequently got their reference number 'wrong': it was always the last figure of the 7, so the offices corrected the so-called mistake by hand and got all the boys' earnings, payments, records, abilities, etc. mixed up. This state of affairs was only discovered when one boy was falsely accused of being hopelessly in debt! It was always said in the unit, that if one of the boys had been killed, they would certainly have buried the other!

3.6 FRUSTRATION

Up to the end of the last century a contract was firm and inviolate: if a contractor was guilt-less when entering a contract but was prevented from carrying out the obligations he had undertaken, he was in breach of contract and suffered accordingly. This was manifestly an unfair imposition, and it was from this that the 'principle of frustration' grew. If circumstances arose under which a blameless party was no longer able to carry out the obligations he had undertaken when entering a contract, his liability could be removed by the courts. We give a few possible reasons below, but the circumstances are so many that it is more a question of having to consider each individual case than to produce a list of them. The principles of frustration cannot effectively be reduced to a set of rules.

It cannot, of course, be applied to anything but a fully fledged contract already agreed, signed and delivered, and even then to each contract separately. Parties must be innocent of any connivance or scheme of attempted frustration. In some of its aspects, frustration is difficult to distinguish from 'mistakes' (of the legal kind), but the best differentiation is that, in most cases, frustration only arises after a contract has been signed by parties which were unaware of any difficulties when they contracted together. They were both innocent.

We give below a number of typical cases but are aware that there are many others which do not fall into these categories:

3.6.1 Non-existence

Unknown to either party (and the more so if one is an agent at some remote place), the subject of the contract no longer exists. It may have been irreplaceably destroyed; for example, it might be:

- damaged or destroyed by fire, flood, accident, etc.;
- forcibly sold by court order, compulsory purchase;
- already disposed of by another agent or party elsewhere;
- buried by landslides, earthquakes or at sea (e.g. as jetsam) or marine disaster;
- lawfully destroyed by an appropriate authority, unaware that a contract of sale was pending;

- unlawfully seized by vandals, guerrillas, insurgents, rioters, foreign states, aggressors, etc.;
- rotted, perished or irreparably deteriorated, or by disintegration;
- ownership – the party concerned may not have the right to dispose of the item, though unaware of it at the time – e.g. an owner may discover his title is faulty, or an agent may be dismissed by a communication from his principal, on its way to him.

3.6.2 Quantity

A seller has contracted to supply a stated quantity, but may find he cannot do so as the contract progresses. The items might be antiques or historic objects and cannot be repeated. The original quantity might have been depleted by vandalism, theft, accident, or the like.

A project manager might need to obtain a site, or land in one piece, for his project. Having concluded his contract, it might be found some of the land is no longer available for one reason or another (see above). He has to reject an offer from the seller of a substitute piece of land 'just around the corner': his project must be all in one piece, and part-sites are no use to him. A fragmented site would force him to go elsewhere. All or nothing!

3.6.3 Quality

Here the situation is much more vague than with quantity. It is often considered that matters of quality should be decided by the intending purchaser before he signs any contract. Otherwise he may have the provisions of the Sale of Goods Acts regarding the adequacy of samples, or 'fitness for a described purpose' to help him. It might be a matter of delayed delivery involving excessive deterioration, or rapid disintegration, or of just what the parties agreed to in their contract. The classic example is a contract to buy a painting which both seller and purchaser and their experts agreed was a genuine one, but which subsequently was adjudged by all other experts as a forgery. It was thus of considerably lower quality, though both parties to the contract were happy, had viewed the picture themselves and the terms of the contract were all faithfully observed. It was difficult to substantiate any charge of negligent misrepresentation or fraud in any form.

3.6.4 Time factor

The circumstances of the site or the procrastinations of a third party might prevent an innocent party fulfilling a completion date that he has undertaken to meet. Neither party to the contract is blameworthy.

3.6.5 Death

If a contract is entered with an individual and that individual subsequently dies, then the contract terminates with his death. Moneys due to him are paid to his estate; moneys owed by him to the living party are regarded as expenses to be met by the estate before probate is granted.

Project managers might come across this situation: contracts with a local man and contracts of employment, and some agency contracts, all come within the category of individual contracts. Contracts made with a firm are usually not affected by the death of an individual, apart from any delay in their appointing a successor. This would also apply to a contract made with an individual but not in a personal capacity; his responsibilities might be transferred to his executors or nominated representatives (successors) who would inherit his rights and obligations, the contract carrying on as before his death.

3.6.6 Insolvency

An individual declared bankrupt (either voluntarily or by order of the courts) following application by himself or his creditors is no longer permitted to trade. His affairs are taken over by an 'insolvency practitioner' as his trustee in bankruptcy. This official acquires all his 'things in action' but he is at liberty to disclaim any 'onerous property' which can include any unprofitable contract. The insolvency practitioner can, as an alternative, adopt and manage the contract, or he can negotiate a new contract with the Employer. If he should decide to reject the existing contract, he is, of course, in breach of contract and may incur due damages accordingly; but as his main concern is a shortage of cash to pay his debts, he merely acquires one more non-preferential creditor who will normally get little satisfaction from the situation. It is here that we should record that a bankrupt's creditors are not all treated equally. There are five categories of 'preferential creditors', most of which concern money owed to the government such as unpaid taxes, VAT, or Social Security Contributions. They also include sums which the bankrupt owes to occupational pension schemes, employees' wages, employees' expenses already paid out, and the like. Further information on insolvency is given in Section 7.6.

The position is very similar with an insolvent firm of contractors, which may go into liquidation on an order of the court. The firm and its current Board of Directors cease to operate and their place is taken by an 'insolvency practitioner' appointed by the court. The official receiver in bankruptcy may intervene, and after an investigation may obtain an order for the firm to be wound up, liquidated and dissolved. The sole interest of the 'insolvency practitioner' is what he considers best for the company itself and its creditors.

In either case, the project suffers (as indeed it does in all matters of insolvency) and a contract which started off with both parties happy, ends in confusion and dismay. The chances are that the contract was originally entered by the Employer, solely on the recommendation of the project manager. Therein lies the lesson.

For anyone interested in the detailed procedures adopted following insolvency, the daunting Insolvency Act 1987, to which we refer in Section 7.6, is brought to notice. It is a fearsome Act, hardly one which can be recommended for easy reading!

4 *Agents and agencies*

4.1 WHAT IS A LEGAL 'AGENT'?

An agency is a relationship between a person in authority and another person whom he appoints as his representative. At law a 'person' also includes a legal organization, duly appointed as such by virtue of a law which gives it its authority. Thus a limited company is given a Board of Directors by its shareholders who, in turn, appoint certain full-time representatives to act in their name and under their guidance – e.g. the managing director, the company secretary, a chief accountant and others. These are all agents of the Board, and derive their authority either from the company's Memorandum of Association (which every company has to produce at the time of registration under the Companies Act or from separate warrants of authority given to the different appointees (or some of both).

Thus the chief executives can be assumed by a project manager to have the powers normally allotted by a company of its size and importance to these executives: anything further restricting their powers is taken to be an internal matter between the Board and the executive and does not have to be investigated more closely by a contracting party. A project manager is entitled to assume they have normal powers, until informed otherwise.

In the same way, the chief executives on the permanent staff of a public authority, duly elected to represent the people under the appropriate Act of Parliament, are the agents of the elected representatives in plenary session. They may (and usually are) empowered to take contractual action on behalf of their masters, but the latter usually control the spending of public funds, by a confirmation of actions of their executives in the form of a sealed agreement by the elected body in session.

There are many less obvious examples of an agency: for example, a partner is the agent of his partnership. But in particular, an agent does *not* include the local (or self-styled) agent who acts as a tradesman exclusively on his own account. We frequently come across this genre in the guise of 'agent for so-and-so's lawnmowers', meaning that they have concentrated on this one brand, carry stocks of its products and possibly spare-parts, and have become recognized locally as the place where this particular product can be obtained. They may even be an appointed 'stockist' and looked on with some favour by the producer of the item but, in every instance, they buy on their own behalf alone. They are not legally 'agents' with powers to act on behalf of the principals. Their having the items and spares is not a part of the producers' overall plan. They sell them to whomsoever they can later, and hope to put to good account being able to say, 'I've got one in stock!'

4.2 SCOPE OF AGENT'S AUTHORITY

At the start, the word 'agent' appears to be a very simple conception – i.e. an authorized representative. On further examination, it is found to have several different meanings as follows:

- An agent may be given *actual* authority by his principal. Actual authority may be either:
  (a) *express authority* – i.e. the principal limits the scope of an agent by specifying in precise words how far his authority extends; or
  (b) *implied authority* – i.e. the agent's scope is implied by the appointment he is given by his principal; a managing director, for example, may not have his scope specified, but a project manager is legally entitled to assume anyone so appointed has all the powers allotted by custom to go with his position in a company of appropriate size.
- An agent may be given *apparent* authority if it appears to everybody that he has been given a certain scope, even if he has in fact been given less. Thus a managing director may be limited by his Board to incurring costs up to £500 only without their prior authority. They tell none of the clients of this limited scope: clients may assume the full apparent authority. The figure of £500 represents his actual authority, but it is only an internal limitation inside the firm.
- A third 'grade' of authority is *usual*. Generally this is just another term for 'implied' or 'actual', or perhaps 'apparent' authority, but it is sometimes used legally to cover some action beyond an agent's 'apparent' authority which, in the circumstances obtaining at the

time, he might reasonably consider coming within the powers of his appointment. It is largely a matter of how far the definition of 'apparent' can be stretched: it is chiefly a legal question of whether he was justified or not.

- Finally, we have the authority of *necessity*. In extenuating circumstances, an agent (out of touch with his principal for prior authority) might assume it, and go ahead on his own responsibility in spite of his being outside his apparent authority or his actual authority from his principal. It may appear to him to be advantageous to his principal (and doubtless to himself!) in the situation existing. The principal is able later to ratify the action and adopt any contract resulting. Or he can refuse to do so, leaving the agent to shoulder the responsibilities undertaken on his own. Clearly, the agent will be convinced of the pressing need for his actions before undertaking them. In these days of improving world communications, the absence of discussions with his superiors gets less and less, and the use of authority of 'necessity' diminishes the whole time.

The credentials of an agent are very important to him to avoid his personally taking liabilities, intended for his principal. His actual authority should exist in writing, as fully and clearly as circumstances allow. We have already seen (in Section 2.3) how an agent must sign any contract on behalf of his principal in at least two places, to avoid personal liability (once in the preamble, and a second time at the point of signature). It must also be clearly specified just how far he may go on his own, without prior confirmation from his principal.

4.3 AUTHORITY FOR AGENCIES

Authority is usually given to an agent in one of the following ways. In general, the document and certified copies thereof, are given to the agent as his credentials, for production to customers on first dealing with them.

4.3.1 By a simple letter from an executive of the organization

This, the simplest form of authorization, is often used when a person holding an appointment (e.g. a managing director), and having certain powers associated with that appointment, wishes to empower another person (named in the letter) temporarily to carry out a series of actions, or single action falling within his own powers. Clearly, he cannot empower somebody else to carry out actions over and beyond his own limitations.

He might, for instance, transfer to an individual who is not normally given such powers by his position in the organization the power to sign a simple contract in his place (the value of the contract lying within his own powers of signature). His letter of authority would form the introduction and credentials to the second party to the contract.

4.3.2 By a more formal and legal document

A body of persons may have wide powers by virtue of their election by the public (for example, a Board of Directors duly elected by the shareholders). The extent of such powers is carefully chosen and laid down in the Articles of Association of the organization. In effect, they *are* the organization, and answering only to their electors, omnipotent. The Board may decide as a whole, to allot some or all of its powers temporarily elsewhere. Usually it will be only some of its powers, and for only a specified and limited purpose, when it is not practicable for them to attend to the specified purpose personally, as one body, for example, overseas. They may legally delegate their powers in accordance with their Articles of Association, in one of several ways:

(a) By all members individually signing the document of authority. This could clearly be an extensive undertaking, especially in remote places.
(b) By adopting a resolution accordingly, at their normal meeting. This is minuted and approved by the Board, and their secretary can give a copy to the agent as his credentials; see Appendix 10 for a typical example.
(c) By issuing a formal document under its corporate seal to indicate it was agreed by the Board as a whole. Such a powerful and formal authorization is normally reserved for a special purpose. It is legally drafted, of temporary validity, and as it gives the bearer a free rein within the limits it imposes, it is usually given to senior members of the organization dealing with contracts or matters of considerable importance. The Board is, after all, temporarily passing its own powers and the responsibilities given to it by its electors on to somebody else. That person can commit his company in any way he thinks fit, subject to any limitations as are in the document itself, legal, political, economic or commercial.

At this point, we might refer to the phrase 'Power of Attorney'. Whatever its use in the USA when referring to a legal practitioner, its derivation is from the French 'attorné' and the Latin 'attornare – i.e. a person who acts legally for another. Such 'Power of action for another' is usually given by a deed, the 'Warrant of Attorney' (as opposed to the other simpler document, the 'Letter of Authority' from, for example, a managing director). The deed bears the firm's seal representing the Board as a body; by this means, they give their assignment of power, and any group of duly responsible people would normally act in this way. The phrase 'Power of Attorney' is (as the name indicates) the actual *powers* which the authorization bestows and not the document itself, the Warrant of Attorney. In spite of the legal rule that an agent is *personally* bound by any form of contract, he might enter *under seal* (even though he seals it on behalf of his principal), this does not apply to powers of attorney (see section 7 of the Powers of Attorney Act 1971).

4.3.3 By a form of agency contract

This is the most usual method a project manager is liable to meet. It is a formal contract drawn up between two firms in which one appoints the other as its representative and the second duly accepts. The principal is prepared to support the other by financial or other means. The contract is binding on both firms and gives details of what services are to be provided, what support is to be given, how local orders are to be handled and, most important, if and how the agent is to handle money arising from local sales or contracts, and how he is to account for it to his principal. During the past decade matters of agencies and their powers have become considerably more complex with the acceptance of new EEC Regulations. We deal with this aspect again in Section 4.8, below. Any limits hitherto imposed on a firm by its Memoranda of Association no longer hold good, and unless curtailed by the principal, the activities of an agent can range ever wider.

4.4 A KNOWN-PRINCIPAL AGENT

This is the type of agent whom a project manager is most likely to deal with when managing a project overseas. An agent's relationship with the principal, his local powers and his aims are all clear and 'above-board', so that the project manager is in no doubt exactly where he stands on his contract. His first meeting may well be with the principal in the UK. They know of an organization which, although locally well known, has so far not had sufficient business in the territory to warrant opening its own department there. Instead, provided that the local firm has the requisite status, finance and experience, it is appointed as agent by the home principal, whom it represents as sales promoter in the territory. Customarily the principal will be bound to act exclusively through such an agent and vice versa on all matters covered by their agreement.

The principal will supply his local representative (agent) with suitable quantities of sales and technical literature, thereby enabling him to negotiate authoritatively with all local potential customers. If he obtains an order, the principal may come himself to the area to check and sign the resulting contract, especially if it is large and complex, with detailed credits and terms of payment. Alternatively, if the principal obtains an order at home to be carried out in the territory, he will transfer it to the agent; their mutual financial arrangements and scales of discounts will take care of such a situation. In both cases, the agent acts as the 'eyes and ears' on the ground for the principal, and he can also give the project manager invaluable support using his local knowledge, his ability with local dialects, tribal customs or taboos. There are many local matters which a project manager must know and take into account when devising his schemes for the project and his programmes for local work.

As regards executive command of the work under the contract, the local agent may appoint a representative to take charge of it for his principal, thereby obtaining full advantages from such local knowledge as he possesses. If such executive responsibility is not compatible with the experience of the agency, the principal may decide to run the contract from home and send out a team of his own (on the lines we describe elsewhere) to carry out site-management. They would then liaise with the local agency. From our own project manager's point of view, an agent is a very useful friend in what would otherwise be a strange and remote country; he should never be allowed to regard himself as purely a contractor's man.

A matter which has to be settled clearly and from the outset is the question as to whom payment of the contract price is made. The whole of the project manager's administrative work will revolve around the technical checking and periodic paying for work carried out, and has to remain unsettled until the most appropriate channels have been decided. There is, of course, no doubt about the legal requirements of the contract, or who is the second party with whom the Employer has signed his contract. The project manager will be in no doubt about whom legally is to be paid, but the practical method may not follow the strictly legal one. In any large local contract, the agent will probably have extensive expenditure in the district for such requirements as vehicle hire, local labour, indigent materials, and so on, and these will have to be paid for in local currency. Much will depend on the arrangements existing between the principal and his agent regarding the handling of money, but to avoid having to transfer sterling into the territory, the principal may authorize the Employer (and through him, his local man-in-charge) to hand over certain funds to the agent in local currency. They will have their own arrangements for accounting of this money and for passing any balance back to the principal in the UK. To some extent, these decisions may rest on local monetary laws, and the difficulties of getting foreign currency in or out of the territory concerned; essentially they will affect the Employer's accountants rather than the project manager, but at long range, the latter might find himself much involved in making the appropriate arrangements.

Equally important from the project manager's point of view will be the question of whom he must rely on if he encounters delay or procrastination from the agency's local representative at the site. Is the local manager in complete control or must any complaint be referred direct to the principal in the UK (with whom the contract is probably made anyway)? Management in isolated outposts can be a very lonely and personal affair and such a procedure might easily result in a decidedly more difficult situation *vis-à-vis* the agent. Alienation of the local authorities must not be overlooked, and the utmost persuasion is tried at local level before a formal complaint is made direct to the principal. It may take longer and mean much more work for the hard-pressed management team, but it maintains a personal touch!

4.5 THE AGENT WITH AN 'UNDISCLOSED PRINCIPAL'

There is a somewhat exceptional case which brings further difficulties for the project manager, namely that in which an agent does not disclose to the project manager that he has a contract as agent with an unmentioned principal. His reason might be political, a tax gain or a gain in his stature locally, or (such a case is recorded) because the principal has been a fierce competitor of the Employer; or, of course, it might be that the agent is a scoundrel and unprincipled. The project manager wishes to make a contract at the project he is taking care of in a foreign location; it follows, then, that any contract must be made with the agent himself, acting as a local firm on his own, with no assistance from any parent firm either financially or technically. The project manager is counselled to take advice from local legal professionals as other difficulties may arise in these circumstances.

The status of the agent's firm is relatively unknown, as is his financial stability and technical know-how. The payment of money locally raises questions on the transfer of funds to the territory, local banking resources, rates of exchange, and the rest. There may be restrictions on the import of foreign currencies to the location, but with most overseas countries short of foreign currencies, this is not likely to be a problem these days.

As no contact has been made with the principal, nothing is known of the legal arrangements between him and his 'agent', the terms of their contract or for the handling of the cash (i.e. the contract price). The so-called agent may be acting genuinely on his own account, or he might be fraudulent. It is most unlikely that any principal would have an arrangement with his 'agent' to provide him with assistance and get no return from so doing: it must therefore be assumed by the project manager that whatever the actual arrangements, they will bind the 'agent' to make some financial contribution from the contract. The principal will be disconcerted if he discovers the existence of the contract and nothing materializes. If the work covered is within the ambit of the agreement, the principal has a *legal right to come into the contract* to the surprise of the project manager, who was under the impression that he had a contract with a local firm. The principal can sue and be sued under the project manager's contract (which is hard to reconcile with the doctrine of privity of contracts), as well as any steps he may take against the so-called 'agent' for breaching his agency contract. In other words, the undisclosed principal can make himself a full party to the Employer's contract. His right to step in is, however, subject to certain provisos:

- The agent must have had authority from the principal to enter a contract of the type he did.
- The principal can come in only if the Employer cannot show he had a good reason to deal with the local man and no other. For example, he might be using personal expertise or experience which only the local man has, but the principal does not.
- Any goods dealt with by the contract were previously obtained from the principal, even though they were obtained on his own account by the agent.
- The agent remains fully entitled and liable under the project manager's contract with him, even after the principal has joined in.

This does not mean that the project manager could sue *both* and collect damages twice! He has to elect which party he will thereafter hold liable, and release the other party entirely, even if his damages from the first choice remain unpaid!

In effect, the principal can intervene only if the contract was one which he should himself have undertaken (under his existing agency agreement), had he been told about it by the so-called 'agent' at the proper time. The local man's right to act for himself in other cases is upheld. The project manager must not suffer on legal grounds.

4.6 PAYMENTS OF THE CONTRACT PRICE

It might, at this point, be worth recapitulating the position in which the Employer finds himself when dealing through an agent. In the case of a disclosed principal (the normal state of affairs), the agent drops out entirely as soon as the contract is signed. The only complication is that the project manager must not pay anything from the contract price locally to the agent without the principal's authority: he owes the money to the second party to the contract, and must pay again if any local arrangements go wrong; the Employer and his project manager are otherwise in the clear. If the local man is unreliable, he may refuse to pass on to the principal moneys paid to him. There is admittedly a great temptation when the site is in an inaccessible part of the world and the local man has incurred considerable expenditure from his own funds on the project, for local officials to refuse to operate until they see the colour of the project manager's money – and for local payments to be made to them for the sake of getting the project ahead in a friendly atmosphere.

The case in which the principal is undisclosed is, however, more complicated. The contract, by definition, will have been made with the 'agent' alone, so any payments made to the agent under the contract are legal. If he chooses not to pass on to the principal whatever he owes him, that is a matter for them to settle under the agency agreement, and the Employer is in the clear.

If, however, the undisclosed (but now self-disclosed) principal decides to intervene in the contract as explained above, and his intervention is legally permitted, the general principle is that such intervention should not place the Employer in a worse situation than he would have been in were the agent truly acting on his own. As regards past payments, the Employer is in the clear; and as regards future payments, his contract is now with two people, the agent together with the intervening principal, and he must presumably pay to each what he is owed under the contract price. The legal position may not be entirely clear, but the ex-miserly 'agent' will hardly be in a position to object!

The above has been considered from the viewpoint of the Employer purchasing something from a second party, while the project manager might be faced with the opposite situation and be *selling* something to the others. The legal ramifications are more involved, and beyond mentioning the case, we shall not pursue them here. It is much less likely to happen in the project manager's case, but the possibility of this situation should be appreciated. The arguments are on much the same lines as in the above, but the advice of a local legal professional should be sought before taking any action. The situation is liable to involve land or a service, such as transportation, rather than goods or materials. It may have to be paid for separately from the existing contract, by another 'strictly local' contract.

4.7 THE END OF AN AGENCY

An agency is the same as any other agreement or contract, and it can end in any similar ways. However, there is one extra matter which always needs to be considered, namely that one or more contracts with outside third parties are still in progress and it is not simply a matter of putting an end to all agency matters by a single stroke of the pen. Outside parties are very much concerned with the future of any agency agreement: the Employer and his project manager must always be considered and new arrangements made for their future servicing on existing contracts. The project must not be delayed by agency changeovers.

An agency can therefore come to an end as a result of any of the following:

- Performance – an agency may be set up solely to deal with a specific piece of work. Having completed its task, the agency can be dissolved.
- If the agency is held by an individual, his death, insanity or bankruptcy (forbidding him to trade) all bring the agency to an end. The contract is probably with the principal, who must make suitable arrangements forthwith.
- By mutual agreement between the agency-agreement parties. Their contract might expire, the local situation might alter obviating the

need for an agency. Mutual arrangements would have to include just what is to be done with unfinished contracts by the agent himself, or what new arrangements the principal would make to take over. It might well involve a revised contract between them.

- Insolvency of one of the organizations concerned. The agency is transferred with other assets to the successors or to the appointed trustees. Their possible actions are dealt with elsewhere in this volume, but the project will inevitably suffer in the way it always does in such situations with any contractor.
- Repudiation, breach of agreement, etc., as normal, the other party sueing for damages. Special *ad hoc* arrangements may have to be made by the principal or the local agent to deal with work currently in progress.

It should always be remembered that the Employer has a valid contract with one or other of the parties, usually with the principal. Legally he is in need of no other legal cover as long as his contract is foolproof. He can sue for breach of contract if the party he is tied to lets him down. Theoretically this is straightforward, but in foreign countries, with strange languages and local customs, it may become a long-drawn-out affair in which a local lawyer needs to be involved. The main problem is to avoid undue delay to the project, and the only actions which can be recommended are: (a) to avoid complications from ambiguous documents; (b) to take action as soon as changes are rumoured; and (c) for the project manager, above all, to ensure that all his plans are concerned with reliable firms and their experienced agents. As usual, the project manager carries the responsibility.

4.8 AGREEMENTS WITH SO-CALLED 'AGENCIES'

In the recent past it has been commonplace to refer to anybody who acts on your behalf as an 'agent' – it is almost better to use the vague phrase 'our man in Brussels' because, at law, the word 'Agent' has (as we have already seen) a particular meaning, and many rules and regulations are woven around the conception of an agency and its taxes. Unfortunately, these are not the same in different countries, so that an 'agency agreement' has to be looked at by an expert in the laws of the country concerned. It is not surprising therefore to find that the word is often replaced by another description, less likely to be given such a wide legal meaning as 'agent', and usually chosen so as to illustrate more precisely just what the appointed person is expected to do for you. Thus we have, as examples:

Distributor;
Stockist;
Sales Office/Promoter;
Sales Representative;
Representative;
Repair station;
Spare-Parts Distributor;
Commercial Representative;
Consultant/Adviser; etc.

The situation has become more confused by our being a member state of the EEC. This group has as its basic aim the matters agreed in the Treaty of Rome, that is the member states forget their individual 'authorities' and subscribe to a common dogma of the EEC and its equality area, the Common Market. The intention has always been to permit unrestrained and equal-opportunity trading between the member countries, and laws have been promulgated to ensure this is done. Thus, for example, an exclusive agency in a territory (such as say, Denmark which is a member state) is against freedom of trade in the Common Market and therefore illegal. Everybody must be free to trade anywhere inside the Common Market. An *exclusive* agreement to distribute products within the entire Common Market is also illegal, as is an exclusive agreement to distribute to the whole EEC, plus a non-EEC country (e.g. a world-wide distributor), equally illegal. There are special rules concerning commissions and the sums due when an agency is closed down. EEC law applies to agreements which might affect inter-state trading, and made between (a) two companies both as EEC member states, (b) UK companies and (c) companies which are not EEC members. Fines imposed often far exceed the apparent gravity of the offence.

It is fair to say therefore that no non-professional person, and anyone not in possession of the complete facts, should attempt to draft 'agency' agreements which might in any way involve EEC member countries; leave it to legal specialists, who are also EEC legal specialists. The law may be different for each category and each country, and the unfortunate word, unwittingly used, may bring in its train a claim for more taxes! Any handling of money locally is a danger signal.

5 *Authentication, agreements, bonds, guarantees and warranties*

5.1 AUTHENTICATION OF DOCUMENTS

There are two ways in which documents are commonly authenticated. Each has its own degree of formality, legal status and validity. They are:

- *Under hand.* That is, under the signatures of persons who represent each of the parties to the document.
- *Under seal.* That is, under the corporate seals of each of the parties to the document.

Engineering contracts will be met under both types of authentication, and the project manager must be aware of the reasons for, and advantages of, each type.

5.2 ENGINEERING CONTRACTS AND AGREEMENTS

5.2.1 Contracts under hand

Each party is represented by the signature of one man, who may or may not be known to the other party. The signature is witnessed, but all the witness states is that he saw a person apply the signature concerned. The following points are therefore left unanswered formally by signed contracts:

(a) Is the person who signs the person he says he is?
(b) Does he hold the position in the contracting company that he claims he does?
(c) Has he the powers he claims to represent his Employer!
(d) Has he the authorization to commit his company to take on the responsibilities allotted by the contract? (Especially the firm's money put up by its shareholders?)
(e) What are the limits, if any, imposed on him by his Employers?

Of course every case is not so wide open; many of the signatories will be well known to the other party, and as far as they can judge, the person concerned holds such a ranking position in the company he represents, that he can be assumed to have all necessary powers. That is all they have to assure themselves on: has he got the sort of appointment which would normally carry such powers? They are not required to investigate further to make the signature acceptable (the Local Government Act 1933, section 266(2b), for example: see Appendix 8).

But not so the courts: they take nothing for granted. Unless the two parties agree as common ground the validity of the contract documents, all these points have to be cleared by testimony from the company concerned, and the evidence may be subjected to cross-examination. Even the signing parties may not be prepared to take the risk if the value of the contract is unusually high, and may demand something more definite than a single person's signature. After all, he might be a crook or an embezzler and the risks are high.

5.2.2 Contracts under seal

The corporate seal of a company is, in effect, the combined 'signature' of the company's whole Board of Directors. They are the people who have been elected as a group by the many shareholders to manage the assets of the company in the way the shareholders wish. In an agreement under seal therefore they have all agreed to the 'agreement' and ordered their 'Keeper of the Seal' to remove it from its hiding place and apply it to the document to show their corporate acceptance. No power is delegated, no subordinate is brought in, any question of their own powers is stated in their articles of association, no limits are involved, no personalities or questions of intent or honesty. They *are* the company, and they have all agreed by the necessary majority while in session to accept the contract documents, with all that they imply.

We can see right away why an Agreement under Seal is said to have greater formality than an agreement under hand. We can also see why a second party will be happier if his large contract is agreed to by the Board rather than a, possibly unknown, individual. An Agreement (or Contract) under hand is known as a 'Simple Agreement': one under seal is known as a 'Formal Agreement'. The person who signs an Agreement under Seal is not signing the agreement or the contract itself; he identifies the company seal and certifies that he applied it (or it is applied in his presence) to the document. The contents of the document are authenticated by the company itself, in the separate persons of the elected members of its Board of Directors.

If a party has no corporate seal, he may still achieve the same result by the use of the formula: '. . . signed, sealed and delivered by me [on this date]'. He duly signs the document before a witness and usually sticks a red adhesive token seal at the same place, to draw attention to the document having been formalized under seal.

A document under seal is known as a 'Deed'. It can be varied or discharged by any written agreement between the parties, or by a verbal agreement (provided that this is suitably evidenced) or by the conduct of the parties alone. Such agreement is often in the nature of a separate contract, and in such cases, it must have all the requirements of a contract – i.e. consideration must pass both ways. In most cases, sealed Agreements are, by custom, spelled with a capital initial letter, to distinguish them from the normal process of two persons reaching the same views (i.e. reaching agreement).

Advantages of agreements under seal

- People may feel that a simple contract, signed by an individual, does not fully represent the importance that they themselves attach to their project. Therefore, they insist that it is recorded in a document under seal. The fact that it will thereby show the approval of the Board of the other party will give their contract the greater formality and importance they are seeking. They may require such a sealed document to replace a simple contract after the latter has been signed, in which case it must be very carefully drawn so as not to modify the existing document in any of its terms. The original contract (with all its associated dates) will then remain in force, and its validity from the date of original signing is not altered. A different wording will amend, or even cancel, an earlier contract and form a new and separate contract, replacing it with new start dates, etc. Many of the earlier terms may not have been repeated, so a less effective contract will often result.
- A similar endorsement may be used by authorities responsible for the expenditure of public funds such as county councils. They are elected by legal votes and are usually part-time; they may therefore authorize their full-time senior executive to enter simple contracts on their behalf, over his own signature. These they later adopt by confirming Agreements issued under their corporate seal by orders of a common session of their members. This gives formal approval by the actual people entrusted by the voters with the expenditure of their public money. A civil engineering contract may be entered by a contractor with a County Roads and Bridges Chief Engineer, all signed and

completed, only to be later endorsed by a sealed Agreement with the main organization (who control the money to pay for it). Here is another case where the wording of the confirming Agreement must be very carefully chosen so as not to amend in any way the terms of the original signed contract. An offer for a new Agreement would re-open all negotiations.

• Some matters must, by law, be covered by Agreements under seal. They mostly concern land conveyance, long leases and share transfers, and they are not likely to be met with by project managers in the course of their duties.

• An Agreement is sometimes used to record negotiations on several points following the submission of an offer by a tenderer. Unconditional acceptance of the offer cannot, of course, be used to show agreement between the parties, and the negotiations may have followed what at first sight appeared to be a rejection of the offer, not readily amenable to submission by the Employer as a counter-offer. Recording of the new basis of agreement is made in a formal Agreement: an alternative might be to call for a fresh offer by the contractor embodying all the new variations, from which a new start is made. A similar use of an Agreement might be to highlight the main conditions of an unusually complex contract, in fact the 'essence of the contract'. While drawing attention to them, such an Agreement must be careful not to alter the original contract in any way. Similarly, the Agreement might be used to list all the items of contract documentation.

• A valuable attribute of Agreements under seal is that they extend the application of the Limitation Act (1939) and its various amendments. Under this Act, the period inside which a claim for breach of a simple signed contract can be made is 6 years; with a formal sealed contract, it is extended to 12 years. There has recently been considerable interest in the added protection which a person might expect under the law. It all arises from the difficulty of determining the date from which the two periods are decided. It is logical that any period can only be counted from the time when a fault became manifest to the user, and some years ago the House of Lords decided that, in the case of covered-up foundations, this might not be for years, long after the contractor had left the site! This naturally caused a considerable stir among the contracting and insurance fraternities, and the search for a suitable end to their liability for latent damage has been going on ever since. We devote part of Chapter 8 to this subject later in the book (see Section 8.3).

• At this point, we might refer back to an earlier use of the formal Agreement, when it is used for one purpose or another to replace or augment a simple contract under hand. If it is the intention so to do even before the simple contract has been made, or if the right is reserved by one of the parties to call for such an Agreement, then the facts should be made known at the time of negotiation and a suitable clause included in the contract. The party who is to produce the draft Agreement is named, as is also the period during which that party must do so, or make up its mind to use its option. This then makes it a breach of contract for the abstaining party; or if the clause is so worded, he loses his option to call for a consolidating Agreement. As mentioned above, the Agreement must not alter the terms of the signed contract in any way, and the signed contract then comes into effect at its normal time and remains valid, whether or not it is subsequently ratified by an Agreement. Inclusion of all the suggested details in the contract itself will remove any suggestion that an Agreement has been sought later in order to make use of the advantages it might give the applicant, for example, to use the extra number of years before the Limitations Act 1939 comes into effect.

5.3 BONDS

A Bond is a gratuitous promise made under seal. That is, it is a Deed undertaking or promising something. Such an undertaking to pay money is called a 'common bond', while one undertaking to produce some form of act or event is called a 'bond upon condition'. A bond is then terminated when the final money is actually handed over, or when the promised act or event is occasioned by the promiser. It is then said to be 'discharged' – i.e. completely fulfilled. Simpler or equivalent documents under hand and without the formalities associated with documents under seal also exist, called by other names (such as IOUs or promissory notes or even 'letters, signed and delivered'), and do not have any of the features attributable to Deeds.

Although a person 'contracts' (in common parlance) to do something, a bond is not a contract in the eyes of the law. It differs primarily in as much as only one party undertakes a responsibility, and a 'consideration' only passes one way from him. A 'promise' is, then, what it is; a 'bargain' is what it is not. For example, a wealthy grandfather might undertake to pay university fees for each of his grandchildren, when they severally reach age 18. He might then die. His executors or successors carry out his wishes, and the bond only is discharged when the last payment to the youngest recipient has been paid. Meanwhile it is a legally enforceable document and can be produced to a court of law without further personal intervention or evidencing. One person only makes a promise, and the 'consideration' passes only from him to his beneficiaries. A bond does not have to be counter-signed by the recipient, but merely accepted as being appropriately worded.

5.3.1 Bonds in connection with contracts

Once a contract has been signed, all of its clauses become legally enforceable, and a promise by either of the parties which deals with the essence of the contract is somewhat superfluous. Either it is already a term of the contract, or if it has been overlooked, it may be added only by a collateral contract varying the original one. Both parties have to agree that the change should be made. Thus Bonds will mostly deal with matters cognate to, but not a part of, the contract itself. They will often extend the latter and may be referred to in it.

There are, however, several types of Bond which are sometimes used in connection with contracts and the project manager should be aware of them; they are outlined in the following.

5.3.2 Tender Bond (or Bid Bond)

This is sometimes called for by an Employer at the time he sends out his enquiry calling for bids. It requires any bidder to promise to maintain his offer unaltered until a decision has been made by the Employer, or until the end of a stated validity period (whichever date comes first), and meanwhile to honour his offer by accepting any contract based on it which the Employer might place with him. If the tenderer defaults and is in breach of his bond, he has to pay the sums specified by the Employer (in his enquiry) or stated in the bond. These are the damages estimated by the Employer which he is liable to suffer by the default, at a crucial moment in his project programme.

He might call for such a bond, for example, to ensure a tenderer for an uncommon article of equipment or a unique component or some specialized form of service cannot change his mind, once his offer has been made. It is rarely used in the UK, its place being taken when necessary by rewording the tender draft offer (sent out by the Employer with his enquiry). In its new form it binds the tenderer in the same way, without having to make a separate bond. The new undertaking – being part of the offer – is automatically extinguished, once the contract has been signed; and it is in the meantime a part of an offer open to acceptance by the Employer.

A Tender Bond is more commonly called for overseas. It can be replaced by a Tender Deposit, a sum of money lodged with an Employer with each offer made, and forfeited at the demand of the Employer if the tenderer fails to keep his undertakings. Appeals against an Employer's decision might be difficult, especially in two different countries. If the amount of a Tender Deposit is properly estimated, it can have just as great dissuasive power as a Tender Bond, but with less formality; however, if many tenderers are envisaged, the total sum may become very high. Only serious contenders will remain in the competition: speculative bids are eliminated!

5.3.3 A Bond of Due Performance

By such a Bond, a tenderer backs up his promise to 'punctually, truly and faithfully perform and observe his obligations' under a contract awarded as a result of his offer. He is, of course, obliged to do just that under the contract itself, and apart from putting the fear of God into him, the Bond achieves little else. It is therefore usually called for in the form of a Guarantee by a surety (approved by the Employer) who is prepared to back the contractor in his efforts. This might become most important to an Employer if his selected contractor were to become bankrupt during his execution of the contract. He might still give the Employer a legal right, but if the contractor has no assets, he is no better off. A guarantee from a reliable third party might save him. The amount of the guarantee for the contract would be estimated as the total expenses he might incur if the chosen contractor turned sour on him. He is not permitted to include punitive costs in the value of his bond, even if a court later awards them to him. In effect, a bond is his 'claim'.

The word 'performance' needs to be carefully defined to cover all obligations which the contractor might undertake. It can include the costs of tests, lateness, non-payment of moneys due, or costs, arising from any associated contract or work. 'Lateness' also has different meanings, from being just a little late (often covered by liquidated damages, which the Employer might not get from an impecunious contractor) to long-term lateness or complete repudiation of the contract by the contractor. The results on his project programme and allied contractors, and the Employer's ability to earn income from his project, might all be greatly affected, and claims escalate accordingly.

5.3.4 Repayment Bond

An Employer might agree to assist a contractor (either as one of the terms of his contract or as a separate issue) by paying part of the price as a down-payment on signing the contract. Alternatively, he may agree to pay interim payments from the terms of payment, in advance of receiving their worth in terms of work done or goods supplied. He will thereby be at a financial disadvantage, and he will be justified in requiring the contractor to forfeit such outstanding debts and return them to him if ever the contractor should fail to earn them before a certain date is reached. Consequently he requires the contractor to undertake to return such sums (up to a maximum stated in the bond) by getting him to issue a 'repayment bond' every time he pays out such a sum 'in advance'. The same safeguard might be obtained by a guarantee from a third party if the Employer regards his contractor as a poor financial risk. It is only fair to remark that the project manager has been rather less than perspicacious if his Employer should find himself in such a desperate position!

5.3.5 A Plant-Performance Bond

Here is one of those occasions in which so much depends on just what has been undertaken in the contract. It is usually a contract for the provision of certain plant, and as such the contract might have naturally come to an end when the plant was delivered, installed and set to work. It was then taken over and the period of warranty started. In such a case, the contractor is free of his contract to supply, but even so, the Employer might be interested to know that his plant as installed has certain characteristics of efficiency, of quality, of reliability and of operation in spite of its having met the mechanical specification which was in his contract, and duly checked at the time of taking over (or completion). Tests for such performance can be made only after the plant has run for some time under its normal conditions of operation and been properly run-in. If the Employer has been sufficiently far-sighted, he will have allowed for these tests, and kept back a retention of the contract price to meet any shortcomings discovered by the tests.

If not, he must rely on a performance bond by his contractor or a corresponding guarantee, backing him to the same extent. Retention moneys, held back from payments during the execution of the contract, can be paid to the contractor as soon as his contract to provide has been completed, and the state of the plant has been checked at the time of completion. The performance tests can then be made as a separate activity some months later when the plant has been settled in. Failure of the tests may be a breach of the earlier contract (or they may not), but it is rather late in the day to go much further than to make the contractor put the matter right. The plant obviously performed satisfactorily at its completion tests or it would not have got so far. There are advantages in keeping its exact performance a separate issue, with its own scale of damages for failure to meet the promised figures.

The amount of his damages will represent what the Employer reckons he will have to meet through the plant failing to achieve its specified standard of performance. Low production will mean hours of extra running, poor reliability will mean additional maintenance costs and repair bills, reduced efficiency will mean extra running costs, and so on. Whatever the courts might later find, the Employer cannot include punitive damages in his bond or guarantee.

We deal with performance testing elsewhere in these volumes; here it will suffice it to say that they are difficult tests both to carry out and to specify. Obviously the plant is satisfactory or it would never have been accepted, so what is being looked for is a matter of small percentages. This is not an easy matter to argue convincingly in a court of law; it is quite likely the Employer will settle at that stage for a cover of his extra expenses, and not follow up a claim for punitive damages.

5.4 GUARANTEES AND WARRANTIES

5.4.1 Guarantee

A Guarantee is, in effect, the backing of the party by a third party, such as a contractor by a bank, whose stability is reckoned to be more securely protected by its assets. A guarantee is thus a promise (usually but not necessarily in the form of a bond) by a third party, C, to be answerable to a party to the contract, A, should there be any fault or miscarriage in the performance of a duty which the other party, B, owes to A under their contract. The guarantee sets out the manner and the maximum extent to which party B's efforts are to be underwritten. Note that a guarantee only operates if the party backed in this way actually attempts (but fails) to perform his duty: it is not an easy way out for a first party at his whim or pleasure.

The Employer first sets out in his enquiry exactly what he requires from a contractor before appointing him, by way of a guarantee of his performance of his duties. The contractor, B, makes arrangements accordingly and accepts the provision of a guarantee in his offer to A. He is selected and duly appointed to the contract. Naturally, B has to pay his backer, C, for the facility, and this charge is passed on to the Employer in the offered contract price. The Employer therefore pays more, but at least he has the assurance from B's surety C, that he will get what he has contracted for, or his damages in lieu, even if he has doubts about his contractor's financial strength.

The formal and legal name for the third party, C, is the surety for the contractor, or alternatively the 'guarantor'. Neither the surety nor the party guaranteed will usually enter into any discussions or arguments as to whether the Employer has become entitled to be paid his damages by the surety. They will call for a clean-cut statement of the situation: if it occurs, they pay up, anything else and they do not. The same sort of decision must be presented to them to show how much of the maximum guaranteed is claimed. The decisions are usually to be made by someone who is not himself a beneficiary, and the word of an independent project manager is generally accepted. His role of quasi-mediator between the contractor and the Employer makes his word acceptable to the sureties as one who can be relied on to see fair play done to both parties. The project manager therefore usually states in writing when he considers the terms of the contract are not being fulfilled by the contractor, in the way envisaged in the guarantee, and gives his estimate of the damges the Employer will suffer as a consequence. These facts he gives to the sureties and they pay the damages, up to the maximum sum stated in the guarantee document (see Appendix 12).

To sum up, there are several points to notice concerning a guarantee:

- A guarantee must always be by a third party. It is usually a financial backing by a surety up to a stated maximum sum.
- A guarantee must be 'evidenced in writing' but not necessarily by a sealed Bond. A promise over the signature of a surety is enough.
- Sureties will rarely undertake 'specific performance' (see Section 5.5, below). They usually restrict themselves to payment of a sum of money representing the damages incurred by the beneficiary as a direct result of the prescribed fault in performance of the contract.
- A guarantee is usually tied to a contract as it is originally made. It must therefore be suitably worded such that it does not become inapplicable if the parties to the contract subsequently agree any variations. These might be, for example, to technical details, to terms of payment, to programmes, etc. The guarantee must go on unaffected by them.

5.4.2 Warranty

Often there is a muddle in the mind of the public over the distinction between a guarantee and a warranty. Careless descriptions result. The basic difference is that a guarantee is always made by a third party (of sufficient reliability), whereas a warranty is a promise made by a party about its own behaviour.

It will be appreciated that if the parties have entered a contract, both of them will be already bound legally by its terms, and an additional warranty *on the exact points made by the contract* will really add nothing to the situation. With new subjects, they amount almost to a collateral contract in cases which are effectively a bargain (i.e. when all the legal requirements of a contract can be present). Sometimes they can possibly be included in the main contract as 'implied terms', or not included at all. In both cases a warranty from the contractor might strengthen the Employer's position. For example, the contractor might undertake to have available a full range of spare parts for a stated period of years, or he might agree that his contractual obligation to maintain the works during the first 12 months after take-over (already in the contract itself) will be read as providing all or any spare parts which might be needed, free of charge and without delay. Such items could be made warranties by the contractor and written into bonds or contracts as required by the Employer.

Note that a warranty offers no protection against the contractor's own lack of financial liquidity or bankruptcy. A guarantee does, as long as the surety does not become the bankrupt!

Many so-called 'guarantees' packed with some household utensil or other are in fact not guarantees at all, but warranties. They are distributed by the makers themselves and provide no extra backing from a third party. They are frequently little more than 'window-dressing' and (in the UK, at least) offer no more than the maker is obliged to accept under recent legislation, and frequently bring no additional benefits to the purchaser. A contract maintenance clause (often referred to loosely as the 'guarantee clause') is no more than a warranty by the second party to do just what it says and no more. To many people 'maintenance' includes daily treatment, the rectification of damage due to outside causes and indeed everything that might be needed to keep the machinery in full operation throughout the year. The same 'warranty clause' is (in their view) wrongly titled when it is referred to as the 'maintenance' clause. This is not what it usually says.

Reference must also be made to a second meaning which is legally appended to certain clauses of secondary importance in a contract. We have seen when dealing with breaches of contract terms that these are divided into two classes in the eyes of the law: conditions and warranties. This usage should not be confused with what we have said, above, about warranty documents (refer to Section 1.5). A 'warranty' (by definition) might rank as a 'condition' (in legal language) and rightfully earn its place in a book entitled *The Standard Conditions of Contract*. Who said the Law was clear and unambiguous?

5.5 SPECIFIC PERFORMANCE

'Specific Performance' is an award that a court of law can grant to a plaintiff following the repudiation or faulty fulfilment of a contract by a contractor. It means the fulfilment of the actual aims of the original contract. For obvious reasons, a court order is not often made against the original contractor, but against some parent company or surety under a guarantee who may become liable to see the obligations of the original contractor carried out.

The main difficulty when ordering specific performance is the organization of the work or its constant supervision by the responsible person against whom the award is made. One cannot easily imagine a court of law, or a bank, or some financial institution picking up a partly finished on-site building contract, and becoming instantly experienced enough to supervise its completion. They would have to retain their own building contractor and be responsible for his task; by its nature, this is often open-ended at the time. It is also a problem for the courts or the applicant to say exactly where his late contract would have led, especially if it had a suggestion of development work or experimental trials in its specification. In other words, just what would 'Specific Performance' mean? When is such a court order fulfilled? Even simple building contracts are often not 100% successful first time. An Employer might have had high hopes but not a few fears for the outcome of any such contract, whereas he would be on to a certainty if awarded 'Specific Performance'; he could withhold certification that the court order was fulfilled until the works were completely successful!

The award of 'Specific Performance' is therefore only rarely given, and in its stead a sum of money as damages is by far the more usual remedy. Damages are estimated by the person applying to the courts as being the amount he would have to find to complete the works by some alternative means or to clear away the offending works and start again. He must produce the cost of returning to the same point as he had so far reached by means of a successful contract. He must include, of course, any sums attributable to the delay to his earning power by not having access to the works at the time he had originally foreseen, and the delay occasioned by having to seek out and contract with a firm able to take over the new work. The award of damages in lieu is a much more practicable procedure and has become legally regarded as a more appropriate compensation in the majority of cases. In a few cases, specific performance is relatively easy, and plain monetary compensation scarcely equitable. House purchase might be a case in point and, in general, cases which involve house occupation, or the rights and ownership of land and buildings, are the most suitable subjects for awards of 'Specific Performance'; engineering works, or selling and buying are the least suitable. There is said to be a claim on record where the contract covered the supply of a set of genuine Hepplewhite chairs, and they were stolen. The courts did not award 'Specific Performance' in that case!

As far as is known, the award of 'Specific Performance' is at the discretion of the courts, not a right of the applicant. There appears to be no legal bar against it, but practicability appears to be the key. It is awarded only when money alone appears not to be a fully just and equitable recompense, for example, with a broken contract to transfer ownership of property.

Guarantors usually confine themselves to a promise only of cash compensation up to a given maximum amount. Specific performance may be awarded also, when the contractor at fault is completely owned by a parent company, which has full control of its finances, its management structure, its management staff, and so on. The parent might then take over the subsidiary's contract and see it through, making all and any changes to the subsidiary's methods of operation which it deems necessary. Its reputation may not allow it to let such a subsidiary fail or fade away, with dissatisfied customers left in the lurch. In many cases (i.e. with the subsidiary having no 'substance'), the situation might be foreseen by a perspicacious project manager, and the original contract placed with the parent company, with authority to sublet the contract work to a fully owned subsidiary. With reliable parent companies, an Employer might accept a formal letter from them undertaking that they will not allow one of their subsidiaries to default. This is regarded as an adequate assurance.

5.6 'QUANTUM MERUIT'

This is another expression which is often used in court awards, either as a result of a claim of 'quantum meruit' or when an appellant wins his appeal for damages but is not given the amount claimed, rather a smaller amount 'quantum meruit', meaning 'the amount which is deserved' or 'merited'. It indicates that the court has taken all evidence into consideration, from both parties, and has decided how much should be given to the aggrieved party on the basis of what he is due. A claim for such an award cannot be judged while the contract is still in force. It might well be continued to readjust any balance the court might otherwise think reasonable. The contract must first be stopped, and then a claim made for the courts to consider.

When such a contract has been broken off, either by mutual agreement or by some enforcing situation, it will be found that there is rarely an implicit formula which lays down what should be a fair recompense for all that has happened so far. The situation could be so vague at the time of signature that it would be impossible to reach a decision which was 'fair' in all circumstances. A contractor will claim he has spent more than is reasonable; the Employer says a half-finished job is no earthly use to him and it will take a lot of time and money simply to remove it. The contractor rejoins, 'What about all the stores I have already delivered to site?', and the Employer answers, 'Take it all away again . . . it's no use to me'. And so they go on, until one party or the other appeals to the law to establish what is a fair recompense, allowing for all the moneys which may have been paid to date.

Similar treatment is afforded to contracts which have been discontinued for any reason. No agreement can be reached as to the guilty party or the cost of completing the work, nor the amount of a fair settlement. One party eventually sues the other for the damages as he sees them, and while the courts agree with him as to blame, they may reassess his damages 'quantum meruit'. Even the question of blame may not be 100% clear: the claimant may have himself subscribed to his eventual costs, and a judge will make appropriate deductions from his claim.

5.7 INJUNCTIONS

An injunction is an order made by the courts which commands a defendant either to do some named thing or (more frequently) forbids him to do something. The former is known as a 'mandatory' injunction, and the latter as a 'prohibitory' injunction. Injunctions are made if the court agrees with an application made to them, usually to prevent a crime being committed or to prevent an opponent pre-empting a decision of the courts prior to a case they have brought being heard and ruled upon.

An injunction can be either a 'perpetual' one (i.e. without any time limit being set on its validity) or an 'interlocutory' one (i.e. one granted provisionally to preserve the status quo). The former is given after a court has ruled that a defendant must take an action to ensure a crime is not committed, the latter before the court has ruled on a matter brought before it. As examples, we might quote a mandatory injunction that the party will cease to make obnoxious smells from its premises, or a prohibitory one that the party will not meet or converse with a stated person pending the hearing of a case.

Injunctions are most usually needed to restrain nuisances, defamations or trespasses (i.e. assault or injury to a person, damage, seizure or interference with his goods, or misuse, dispossession or unauthorized entry upon his land). There is no need that any offence has actually taken place; application for an injunction can be made if there is a reasonable certainty of threatened unlawful conduct if one is not granted.

5.8 TORT

A tort is a breach by an individual of a civil duty to his neighbour imposed on him by common or civil law. It has nothing to do with a contract or breaches of contractual terms. The duty might be one of negligence or lack of due care, of respect for a person's possessions, avoidance of damage, and so on. Cases are brought to the attention of the law by the injured person suing the perpetrator (the police are only involved with criminal offences – i.e. those endangering the public at large).

A party to a contract might breach one of its terms by means of a tortious act, and the injured party then has the choice of seeking recompense either by contract law or by common law, in which case the contract is left out of consideration and the party might be left with a faulty contract on his hands. Or he may lay two charges, one to clear his contract and the other to deal with the offence at common law (see Section 3.1).

A master is responsible for torts committed by his servants, but a contractor is not responsible for those of a person he has employed for his special skills (the latter is regarded as a 'master' in his own right – the contractor is regarded as not telling him how to go about his business and he is his own 'boss'). Otherwise a contractor is responsible for seeing that his employees 'behave themselves'.

The law on torts is long and detailed. For example, the rule which excludes persons of unsound mind is only effective in some cases. It also deals with, for example, the difference between a surgical operation and assault on the person, when the person concerned is unconscious!

We merely draw the attention of project managers to its existence, and the possibility of following up the civil injury when it conflicts with a breach of a contractor's contract. The decision as to which course is best followed in any particular case is clearly one for the legal advisers of the Employer. The case may be stronger under one or the other and, in any event, the damages which could be awarded are not the same in the contractual and the common law cases. We do not have to add that a charge under common law does not require that there is any contractual relationship at all.

5.9 'FORCE MAJEURE'

This is a phrase frequently met with in draft contracts, especially those with overseas parties. It is a bigger source of potential trouble than many others, primarily because it has no definition (nationally or internationally). It is sometimes used (like its fellow time-saver 'etcetera') when the writer is too idle to think his problem through and give specific wording: he uses it as a 'sweep-in-everything-else' expression which, at law, has no definition.

It means 'greater force' or by a 'superior being' (Act of God, but what happens if the two parties revere different Gods?). Usually it is applied to mean unexpected events, beyond the control of either party. Neither is to blame in any way. If we were to describe it as an unexpected occurrence, which any prudent contractor would meet by a risk-cost in his original offer, we come up at once against the questions of what constitutes a 'prudent' contractor, and how high a risk should he take without including any contingency, in these days of tight competition?

A wet weekend in the Lake District is a regular occurrence: in the Sahara Desert it is a rarity – an Act of God. A snowstorm or a hurricane (as in the UK, in October 1987) – are they instances of 'force majeure'? Death of a contract manager on the job? A calamity for his project manager, but a 'gift' to a delinquent contractor. Everything depends on what the writer intended, and more often than not, what he doesn't say. 'God' could be replaced by dictators, presidents, revolutionaries, elected governments, armed forces, terrorists and the rest. It can so easily be regarded as a *multum in parvo*, and an excuse for delay, breach of contract, poor quality, lack of planning and lack of foresight.

The expression 'force majeure' must always be clearly defined in the actual contract in which it is used. It must be made precise, not a ground for complicated legal argument. In contracts it should never be used: if it means something, say what it means. If it doesn't, leave it out. An additional clause on the lines of the following might be considered in cases in which 'force majeure' cannot be replaced by something specific that the draftsman has in mind:

Clause XX. Frustration

1. If either party is prevented from, or delayed in, performing any of its obligations under the contract as a result of industrial disputes or of any other circumstances beyond its reasonable control, such obligation shall be suspended for so long and to such extent as may be justified by the circumstances.
2. If work under the contract has substantially ceased as a result of circumstances falling within the provisions of this clause and is not resumed within a reasonable time (for instance, but not limited to, 3 months), either party may, by 3 months' notice given thereafter to the other in writing, terminate the contract.
3. If the contract is so terminated, the parties shall propose and the Employer shall decide and pay to the contractor such proportion of the contract price as the work actually finished bears to the work for which the contract price was fixed, together with such costs as have been necessarily incurred by the contractor in connection with the works so done and the process of termination, provided that they have not been otherwise paid already.

5.10 LICENCES: PATENTS, COPYRIGHTS AND DESIGNS*

A licence is a written authority given by the owner of a right to another person, enabling him to make such use of the right as the licence may state. A licence is usually given for a consideration (a fee or royalty) and is, in effect, a simple contract between the owner and the intending user. A licence may be a 'general' one to 'make such use as he thinks fit', or may be limited to:

- making or producing a single example (as part of an installation of a project);
- making or producing a series of single examples at different sites, or at the same site;
- making or producing a number of examples in a works (i.e. production or mass-production);
- making or producing and selling at home (except for named competitors, for example);
- disposing by sale of a single example as part of an installation;

- marketing and selling overseas (in such territories as are affected by the need to license);
- modifying the subject or extending its field-of-use by development or experimentation.

There are four main ways in which an owner may protect his rights of ownership:*

- Patents on inventions.
- Registered designs.
- Copyrights.
- Deeds of Ownership.

The most likely means to be met with by a project manager is that of a patent, granted in respect of a subassembly or a process which forms an important part of the Employer's project. There is no problem, of course, if the patent is held by the Employer's firm: his licence to the contractor to construct is implicit in his order so to do. If the contractor owns the patent, he must give the Employer a licence to use the patented matter in his project, and eventually to dispose of the same when it is no longer required. The main problems arise when the patented process is held by a third party, and it is on this aspect that we shall concentrate in what follows.

5.10.1 Patents*

A patent is granted to an applicant in two stages, a patent application (which is awarded as soon as the invention is lodged at the Patent Office in London, and they have assembled prima-facie evidence that it actually represents a new invention). Protection is thereby given to the applicant from the date of lodgement, for a certain number of months, with the right to use the phrase 'Patent Applied For'.

The second, more expensive, stage follows after a period of time (at present it is some years) and the award of final Letters Patent may or may not be taken up by the applicant. It is the formal award, with patent-date, official publication, serial patent-number and final claims and specification. The document is usually an extended form of the application, with numerous manifestations of the invention enumerated and claimed separately. All relevant applications of the invention are set out in some detail, with a full description of the methods proposed for each. The document is usually drawn up by patent attorneys

* Since section 5.10 was written and released for publication the Copyright, Designs and Patents Act 1988 has become law (see Appendix 14).

for the applicant, and follows a detailed examination of the application by the Patent Office searchers, to ensure that only genuine inventions are involved. Special rules then cover persons adopting the subject of the patent during the intervening period since the application date. The phrases 'Made under Patent No . . .' or 'Made under licence from the patentee' can be used. Licences are not required for use under stage 1, only after the detailed examination and stage 2 have shown that a true case of invention has been proved and the Letters Patent actually issued.

A patent cannot be taken out on an *idea* alone: if this were allowed, there would be no end to the far-fetched, featherbrained schemes which would be put forward. Every idea must therefore be accompanied by at least one method of application, so that its practical use can be demonstrated. There is no requirement to show economic feasibility, but merely to back up the basic idea with proof that it is a practical one, capable of being put into effect. The wording is often (but not necessarily) on the following lines: 'It is already known that such-and-such are usually produced by such a process. This has many disadvantages such as. . . The present invention is aimed at overcoming these, and this is how it does it. An extra step is introduced into the process by an assembly made in accordance with the invention and illustrated in the diagrams at Figs 1–4. The resultant product now has . . .', and so on.

The enemy of all patents is, of course, proof that it does not refer to an original invention. In official parlance there are 'prior publication' or 'prior art' available to show this.

Just what constitutes 'prior publication' is carefully defined, and roughly it covers convincing proof that the idea was made available to the public at a date prior to the application date by the present applicant. It is not required to prove whether or not the applicant actually saw or was aware of the item quoted; merely to show that he could have done when it was published for the public to see. The writer of the article concerned had clearly had the same idea at an earlier date. A particular case occurred when a party wished to use an existing patent in each of the items it sold. They discovered that the same idea had appeared in a somewhat obscure technical journal from the USA which had been laid on the table at the public library attached to the Science Museum, London, one day prior to the date of original application of the patent. The latter was therefore void, even though the patent examiners had not come across this American prior art when granting the full patent. The firm therefore took no action towards licensing, and added the knowledge to their armoury in case the patentees eventually decided to sue them for infringement. Next move to you!

There is in all this a warning to which the attention of project managers and their staffs is drawn. They should never listen to tenderers who ask to discuss their latest invention 'in confidence' before including the scheme in their tender. If the firm concerned has any plans to protect themselves (which the negotiator may not be aware of), they may be thwarted by prior publication, and there will always remain the possibility that disclosure resulted from a breach of confidence by the project manager's staff or his Employer. Both would come under suspicion as the initiators of the chain that led to the prior publication, and the reputation of both is 'on the line'. The negotiators must be told: 'Go away, and come back after you have lodged your patent application. It costs but little, and only takes a day or so, but you are thereafter protected as from the application date, and can bare your thoughts to us with impunity.' The loss of a few days, and the possibility that a fee or royalty may have to be paid, are better than the potential loss of reputation for passing on matters told in confidence. Indeed the positive advice the other way is often much appreciated by the tendering firm.

A 'fee' is a lump-sum for a licence, and is suited to the use of one example of the invention as part of a process. A 'royalty' is a payment 'per piece' for all units made with or treated by the invention. The selection is at the choice of the inventor, but naturally the Employer will opt for the one-off payment whenever it is at all appropriate. Licences are not normally required until the full patent (i.e. the second stage) is approved. A decision not to proceed beyond the application stage has to be made very early on and, in any case, the Patent Office examiners may find prior art and not award full Letters Patent.

When a licence is sought, the owner of the process must be approached on two fronts, as indicated above:

- The Employer seeks a licence to use the patented process and to resell the licensed plant if it is no longer needed.
- The contractor (or the Employer on his behalf) needs a licence to build the process including the patented plant.

Considerable care is required with both points of view, to ensure the two parties have adequate information or indemnification through guarantees, lest the patented process fails or the owners come into the hands of official receivers in bankruptcy. In either case, the parties risk being left without remedy. There is an additional risk, that the inventor may die or renege, prior to completing the transfer of details to them.

5.10.2 Registered designs*

Registration of the design of a product (for example, its shape and size of container) or of a firm's trade mark (or logo), or of a finished piece itself, may be undertaken, in order to avoid production or use of duplicates by other manufacturers. Note that it is not the *idea* that is registered, but the manifestation of it, so that an object which does the same thing exactly but has a different appearance is not affected. The sole aim is to prevent unauthorized people passing off their product as one from a valued predecessor in the field.

A design is accepted for registration only after the applicant has convinced the authorities that the step introduces no chance itself of introducing misrepresentation. This can be a long and tedious process, especially if any 'wording' is involved in the new design. A similar 'right' to an appearance can be achieved purely by the passage of time and a continued usage throughout by the applicant. Eventually it becomes associated only with the applicant's product, and its value to that firm is appreciated by the registration authorities. Until such time is reached, of course, the firm in question has no rights to the design and can expect to have imitators. Having said this, a design which may not be accepted for registration right away may on its own merits and continued usage, eventually become the property of the firm which perpetuated it.

5.10.3 Copyright*

Here the intention is much the same as that above: to prevent a person taking 'lumps' wholesale from an existing publication by another author, and passing them off as his own, or using a pictorial design to which he has no rights. Note that copyright does not cover the use of an idea, but the actual use or layout of words and phrases. If one alters these, one can write about the same subject-matter: it is no longer a 'copy' of the earlier work for which the author's licence is required before use.

A project management team is most likely to face questions of copyright in connection with computer programs*, a comparatively new issue in which constant developments and changes occur and give rise to new interpretations of the copyright regulations. A written exposition of a program is certainly a 'written document' coming under the copyright regime. This embraces the so-called 'running-list' of a program.

* Since section 5.10 was written and released for publication the Copyright, Designs and Patents Act 1988 has become law (see Appendix 14).

Copyright law is not solely concerned with documents actually written in the English (or any other) language; it has long since been established that it also embraces documents in codes, symbols, and the like. It has therefore been ruled that a computer-based program set out line by line on a dot-code (or a magnetized dot-code) also comes within the copyright laws*, and by extension any other computer program in the form of a separate record. But what about a computer-control system, which was bought complete by the Employer, with its operating program permanently in its ROM. The original program was perhaps not written by the firm supplying the computer system, but by a private writer. The computer firm had somehow licensed it, but had the writer at any stage sold them the copyright? What were the terms of his licence? Presumably the computer-maker had paid for the sale of the program, and was entitled to let the project manager (and the Employer) use it. But could they sell it? Should they pay a fee or royalties to the writer? Are they liable whenever they use his program? It is not the *idea* that is under consideration, but the copyright in the original program.

An architect once drew up plans for a wealthy client, who used them to build a large country house and a wing. Many years later the client added a second wing, and employed a second builder to do it, using his plans of the first wing. Even though the Architect was not called in, he claimed his normal fees for the use of the plans again: he claimed the copyright still vested in him, even though the client had paid for the plans he supplied. The courts supported the Architect. The situation is closely akin to that of a photographer, a professional who took posed pictures at his studio and charged his clients large sums for the prints they ordered. These charges do not include the copyright, which still remains his, unless the client expressly asks for it at the time and pays such extra charges as the photographer levies. The client buys the copyright as a separate contractual undertaking. These two examples illustrate the principle that the copyright vests in the writer (or composer), unless and until the copyright is sold by him expressly to somebody else. Mere paying for the work leading to his writing does not transfer the copyright.

The same principle probably applies to our computer programme-writer. Did he sell his copyright to his client? Or did he merely include a copy and a licence to use it? Did the licence include selling the copy to the Employer, and the Employer using it on his project? Can the Employer *resell* the computer system when he is finished with it? Or can the Employer be sued by the writer for breach of copyright? The ideas behind the system are another matter. They may include original inventions, but these are only protected if they have been awarded Letters Patent (or are temporarily protected while the application is being examined at the Patent Office). If the inventor was also the supplying firm, the Employer is automatically authorized to use the invention for his expressed purpose by dint of purchasing the equipment complete, but he is probably not entitled to resell it to an unspecified purchaser for an unspecified purpose.

You cannot let anyone copy your photograph from one of the prints without paying the photographer for use of the copyright. If he still owns it!

5.10.4 Ownership

For the sake of completeness, a few words need to be said on the subject of ownership. It is really open to doubt whether such a subject should be considered alongside such things as patents and copyright. Everything which is not owned by the public at large or dedicated to the public service is owned by somebody, and they all presumably have some evidence to show that it is theirs. In this way, they protect their rights of ownership. Nobody else has the right to use the thing, and to make use of it they must get from the owners an authorization so to do. This amounts to the owner's licence.

Thus they may obtain the right to fish in private waters, hunt in private woods, use a hired car or a television set, run lorries across private fields, draw water from private sources, work privately owned quarries, camp on private land and many other similar activities, some of which may well come the way of a project manager. For such use a 'licence' is needed and a 'rent' or 'fee' is paid. In some cases, this may be a full economic one, in some a 'pepper-corn' rent (to support a simple contract by passing a 'consideration' both ways), or in others on the say-so of the owner (i.e. no written contract exists to which the project manager can refer, or on which he can rely if the worst comes to the worst).

Clearly, the field is very wide, and we are not able to branch off into a dissertation on the law as it affects ownership. All one can do is to issue a word of warning, that everything (or almost everything) belongs to somebody, and a licence of some sort must be obtained before it can be used by a project manager. This should be obtained in writing, and a receipt obtained for any fee or rental paid (giving the date and the period which the payment covers). The owner may or may not be an individual: it might prove to be a national body (such as landing an aeroplane at Heathrow Airport, where the owner is the Airport Authority and the licence to land is the landing-fee prescribed). Somebody still owns the site and their authority must be obtained before it can be used.

The interplay of ownership and legal trespass is a complicated one which we shall not embark on here. Clearly, any licence or authority to use, as of right, an ownership by another removes any threat of proceedings for trespass.

6 *Exports: shipments and payments*

6.1 EXPORTS

Although the principles of contract are much the same overseas, there are many complications, which although they might not be a direct responsibility of the project manager, may well be connected with his project. He might be in charge of a site situated overseas, or equally his project at home might involve a purchase of goods from overseas which have to be imported into his country. He must therefore be aware of what the procedures are and be familiar with the terms used in every-day dealings with exporting and importing. Thus far will we go in the following chapter. Questions of trading abroad and shipments overseas would comfortably fill a whole volume on their own and indeed have already done so and anyone concerned with such matters is referred to these comprehensive works on the subject. Our treatment will merely touch the practical surface, and introduce the terms which are peculiar to it.

We are told that exports are essential to the well-being of the UK economy. Inevitably, however, they introduce complications for any firm accustomed to the every-day problems of home contracting; for example:

- A contract may be in accordance with a foreign legal system and not related to English law. Local advisers become essential.
- At least one further government is involved, that of the overseas country. It may introduce laws, rules and regulations over which the exporter has no control.
- At least one additional main contractor is concerned – i.e. the carrier – and he becomes the detached possessor of the contract goods for a period of time.
- The buyer and the seller are no longer (as firms) of the same nationality, and are subject to different laws, customs, regulations, etc., and

probably different languages, with all the difficulties of reaching agreement which they introduce.

- Payments take a long time. They may go through a complex financial chain. They may be owed by a firm of which little is known, whose creditability is not known, and who can readily withhold moneys after a process of payment has been started.
- A variable rate of exchange may be involved. The currency of the other's country may not be acceptable or of doubtful value. Various licences may be needed.

And so we can go on. The differences are legion, but as a nation we are still required to export, and various means have been devised to prevent such normal difficulties reflecting adversely on the activities of any firm which takes them on.

There is a bewildering array of documentation which is needed, and a regular procedure to be adopted if the full obligations of the original contract to buy and sell are to be enforced. Firms which export regularly are recommended to make use of the services of an export agent, who can be well versed in the intricacies of the system, and can relieve the exporting firm of much of the paper-work involved in arranging transportation and a succession of payments.

6.2 METHODS OF TRANSPORTATION

Until a few years ago, there was only one method commonly available to engineering contractors in this country doing business with buyers overseas. That was by ship, from a UK port to the most convenient port for the buyer. The goods had to be carefully packed against the rough handling they received during shipment, protected against continued exposure to salt atmosphere, a long sea voyage (usually in the hold of a small cargo boat), unlimited storage in the open air or in warehouses where they could be attacked by local vermin or insects.

Nowadays there are alternatives. First, we have air freight from an airport in the UK to another airport convenient to the buyer's site. The latter may well be far from a coastal port and when in an otherwise inaccessible spot may well have a new airstrip developed for the purposes of the project alone. The periods of the journey and any local storage may be only a matter of hours; packing is much lighter and less comprehensive; and handling is minimal and professional. But of course, the size of any load is strictly limited. Second (and more especially in countries not far distant), we have direct delivery by truck, using the roll-on/roll-off facilities now available on most ferry routes to and from the continent. A number of new features can be summarized as follows:

- Timber and other materials for heavy packing have become excruciatingly expensive and scarce. Making and fitting of packing cases is slow.
- Air transport and charter aircraft are here to stay and daily become easier to use.
- Fork-lift trucks and palleting are now standard practice. Shipment of small consignments can be sent as part of sealed container-loads.
- Airports are multiplying in all less well developed countries, especially when remote from the coast. Import laws and customs facilities have been amended to suit them.
- The introduction of TIR (Section 6.3) and other international agreements in conjunction with roll-on ferries have facilitated direct door-to-door methods of transportation.
- Frequency of services to more distant parts is much greater when new methods of transportation are used.
- Engineering equipment has become increasingly lighter, more electronic, automated and more vulnerable to damage. Value-to-weight ratio has increased dramatically, and with it insurance costs.

Instead of ships alone, one now has to make a selection of the means of transportation for each load, destination and occasion; a balance needs to be drawn between:

- cost of packing required by sea, air and land.
- cost of freight;
- location of port, airport, etc. relative to site; the nature of access between them; and load-carrying capability;
- speed of delivery;
- relative insurance costs; risk of delay due to any damage to cargo; and replacement facilities;
- distance and location of destination from the factory; and local repair and maintenance facilities;
- type of equipment; vulnerability; and size and minimum package weight.

Perhaps a few examples of the wide choice available nowadays might be given. An aircraft was chartered to carry delicate control-room instrumentation from an airfield in North London direct to within a mile of the site in southern Sweden. The instruments were packed in light cardboard boxes, each weighed a few kilograms, and were loaded on to the plane by our own people, who accompanied the consignment (with no fares to pay). Damage was nil. Insurance was extremely low, in spite of the high intrinsic value of the cargo. A similar exercise to Australia was impracticable. The crossing of the Indian Ocean meant chartering a 4-engined plane, the weight of the cargo being insignificant compared with the plane fuel which had to be carried. Economically our load could not fill such a plane.

On the other hand, an electronic computer in a cast-steel case intended for use aboard naval submarines came to grief on being carried by sea to India. Its wooden packing-case was undamaged externally, as was the steel computer case, but the latter had torn away from its holding-down bolts, and rattled about inside the packing. Damage internally to its electronic circuits was extensive and it took a full 6 months to build a replacement (then sent by air!).

Using a suitable large trailer body and TIR facilities, light engineering equipment has been sent door-to-door in East European countries fixed to pallets with little or no extra packing. Fixing was by nylon ropes, and no trouble or damage was experienced. Delivery was of a few days, and included a rest-period over a weekend *en route*. The most difficult arrangement was the convenience of the customs officials concerned! The TIR system is equally well suited to the transport of floor-standing, rack-mounted or control panels which can stand on the floor of the truck and be held back to its top framework, with perhaps a little padding but no extra packing.

6.3 TIR FACILITIES

TIR is the 'Transports Internationaux Routiers: The object of these internationally recognized facilities is to minimize the formalities at frontier stations when a vehicle and its load has to pass through an independent country *en route* to its destination. None of the load is for delivery in the intermediate country itself. The principle is that the load is sealed into its vehicle body and the only entry formality is the provision of paper work (notably invoices) which describe the load passing through the country. Its purpose is to give written assurance that the load contains no materials which are forbidden by the country concerned, for example, arms, drugs, explosives, or the like. On leaving the country, the seals are examined to ensure they have not been opened or tampered with while in the intermediate country. The effect is that a consignment is checked aboard the vehicle against an invoice by a customs official at the point of departure (i.e. the seller's premises), only to remain unbroken until removed by a customs official of the country of destination, and the contents unloaded against the accompanying invoice (i.e. at the buyer's site).

The vehicle body, or the separate container as the case may be, must be approved for TIR work, that is it can be effectively sealed. This cannot be circumvented without leaving cuts or other marks to alert any examining customs officials. Canvas awnings, suitably made fast at the edges are acceptable. An approved vehicle is allowed to carry a square blue TIR plate with the initials in white capitals. Frontier paperwork is limited to the vehicle itself and its contents. In our personal experience, when the system has been used, it worked very satisfactorily, physical searching or unloading at intermediate frontiers was avoided and, in at least one case, the towing engine was changed, only the trailer being sealed and separated.

The sealed container must leave a country within a specified period after entry: the interval is very reasonable and, in practice, the time in the country is only a matter of a few hours. Vehicle crews are treated as through travellers and require valid passports in the usual way. It is understood that the idea was first instituted to permit southern European countries to bring fresh fruit and vegetables to London, without noticeable deterioration, ready to go on to the London market as fresh picked. It is now quite commonly applied and well known to the customs authorities of the various frontiers.

Its usefulness to exporters is great as long as main trunk-roads can be used. If the final destination is in some remote area served by small roads or tracks, the average TIR vehicle is unsuited to such routes (designed more for fast movement along trunk roads) and a vehicle change has to be undertaken at some central point in the destination country, where the seals are broken and the TIR load received by the customs official.

Invoices are required more as a list of the goods sealed in and their value than as a document to be paid, the buyer's liabilities in this direction being fixed by his contract terms rather than the contents of a single truck.

6.4 INCOTERMS 1980: CONTRACTUAL REQUIREMENTS

In framing any contract, it is always important to specify exactly the point at which property in any goods purchased passes from the seller to the buyer, and at what point the goods are delivered. These questions become even more important when physical transportation brings in a third party, the shipper, carrier or his agent, and when the stages and their associated documentation are both complicated and expensive. There are, when we come to analyse the process, a number of distinct phases between seller and buyer:

- Packing for transportation; and storage at seller's premises.
- Loading on to rail or truck.
- Movement by rail or road to port or airport.
- Loading on to ship; and movement by carrier.
- Freight charges; and marine or air insurance.
- Cost of unloading; lighterage; stevedoring; handling on quay; storage in warehouse; and airport charges.
- Importation to country of destination; licences; and other 'port of disembarkation' dues.
- Road transport to site; unloading; and storage until needed on contract works.

The transfer of property could happen before any phase, and this would decide which party had the responsibility for arrangements in the next and following stages. In 1953 the situation was standardized by the International Chamber of Commerce, and they published their agreed INCOTERMS 1953, which gave an agreed definition of each stage and divided up the duties as to what the buyer and seller must do in each case. This publication has achieved international recognition, and subsequent editions have updated the original publication, the latest being in 1980. Most of the methods of transfer of property in the goods are now defined by initials (see Appendix 5) and most of these are covered by INCOTERMS 1980. Naturally both parties prefer not to be responsible for what takes place in the other's country, and it is hardly surprising that the stages FOB (free on board) and C&F (or CIF) are the most popular selections for inclusion in the controlling contract. Each party (by reference to INCOTERMS) then knows exactly what it has to do and where the other steps into the picture.

In FOB, for example, the overseas buyer arranges the ship, the shipping space, pays the freight and insurance, and so on. The UK seller is responsible for getting the goods to the nominated port, and loading them 'over the ship's rail' of the appointed vessel. In CIF, however, the seller is responsible for getting the goods to the far end (though still in the ship's custody), the buyer than takes over, unloads the goods on to the quay, pays the import charges and takes the goods out to his site. The seller consequently pays the freight, and marine insurance and makes all arrangements for the shipping space and the ship. This split is set out in the internationally agreed book, *INCOTERMS 1980*, and is described as such in the contract between buyer and seller. Similar terms are used to describe transportation by air or TIR; and these are employed in the same way.

It is, of course, something of a misstatement that one or the other pays; in the end, it is the buyer who must pay the lot, but in the meantime each party knows just how far he has to go and what he must quote for in his contract price. Although with FOB (INCOTERMS 1980) arrangements, it is the responsibility of the buyer to arrange shipping space and marine insurance of the goods, these are usually arranged in the country of supply by the seller acting in those fields as the agent of the buyer. The latter, of course, pays.

6.5 BILLS OF LADING: AIR CONSIGNMENT NOTES

These are similar documents, depending of course on whether transport by sea or air has been chosen. There are slight differences, so to avoid double explanations every time, we will concentrate on Bills of Lading. They are probably the most important of all shipping documents and act as the official confirmation of consignment instructions to the carrier, being:

- The carrier's formal receipt for the goods, confirming that they are held in custody on board.
- Documentary evidence of the carrier's contract to transport the goods, to a named destination.
- Title of ownership: whoever holds the original Bill of Lading duly signed by the master of the ship is virtually the owner of the goods.
- The document is negotiable: ownership can be transferred without physically moving the goods.

A Bill of Lading (or an Air Consignment Note) usually contains the following particulars:

- Name of shipowner, carrier and the ship itself.
- Name and address of exporter (seller).
- Name and address of consignee (usually 'open order', or the buyer – or his agent) (see below).
- Port from which goods are dispatched.
- Port to which goods are consigned.
- Who pays (or has paid) for the freight charges.
- Description of goods (packaging, marks, numbers of packages, etc.) and accompanying documents.

- Conditions of shipping (e.g. FOB, CIF, hold-cargo, deck-cargo, refrigerated, etc.).

Bills of Lading can be 'clean' or 'dirty'. The former is one on which the master makes no adverse comment on the condition of the goods as received aboard (e.g. packing faulty, inadequate, damaged, package missing, etc.). He is responsible for the safe delivery of the goods and naturally must protect himself against 'as received' damage. Bills of Lading are made out by the ship's master usually in triplicate, each copy being called a 'part'. One copy goes to the buyer (often by the same ship), one stays with the goods and the third is kept by the seller as his acknowledgement that the master has undertaken to take the goods to the port indicated.

The Bill of Lading indicates that the holder is the rightful owner of the goods: the master does not have to determine the right of any person holding the Bill of Lading to receive the goods from him and clear his contract. A Bill of Lading can therefore be used to tie the delivery of the goods and a satisfactory completion by the buyer of the agreed method of payment. If the contract specifies the payment in cash of a certain sum of money, the Bill of Lading is not sent to the buyer until the cash has been received. Similarly, if a prerequisite is that he shall have accepted a stated Bill of Exchange, the local correspondent of the bank representing the seller will not hand over the Bill of Lading until the buyer has actually signed and dated the Bill of Exchange. Only then does he have title to the goods and can get them from the carrier in exchange for the Bill of Lading now sent to him.

In that way, the goods are withheld by the carrier at the point of consignment, depending on how the 'consignee box' is filled in. If the consignee is named as the buyer, it means the master can hand over the goods to him (this assumes he has met the payment clause of his contract with the seller – the Bill of Lading can be sent to him by mail or by the same ship as is bringing the goods). A second method of completing the 'consignee box' is to mark it to order of a bank or of an interested third party. This is used when the party needs to have control over the transactions between the carrier and the buyer. The third party (when satisfied) endorses the Bill of Lading and transfers his interest to the buyer, who can then claim the goods. The third method is to mark the box 'to order and blank endorsed' (usually by entering just the words 'to order' in the box on the form). This is the instruction to the carrier to withhold the goods until they are released following the receipt of payment by the seller, and the shipping documents being passed to the buyer. Similarly, if the delivery is to be on 'open account' or purely on the buyer's credit, he will be sent or given the Bill of Lading on arrival at the port of consignment and will immediately take delivery of the goods. The shipping documents are held, in effect, as security for proper payment by the buyer. The whole system is today too slow for shipments 'on wheels TIR' or 'by air-freight' as the goods can arrive ahead of the papers. Quicker procedures have to be substituted, either by demanding the establishment of a credit at a bank in London to be collected ahead of the dispatch of the goods, or by withholding delivery until the buyer has established a credit with the correspondent locally of the seller's London bank. Or delivery can be made 'against documents' at the receiving end, a form of credit in which the seller trusts the buyer will honour an accepted Bill of Exchange or a cheque or banker's draft or other sight document at the time of delivery. Somebody has to trust somebody else in most of these alternatives, and it is rare for a Bill of Exchange, once accepted, to be dishonoured.

From time to time, project managers might come across another receipt, issued temporarily by the ship's officer in charge of loading and known as a 'Mate's Receipt'. This is merely a temporary receipt given to the exporter to say that the goods have been duly taken on board. It does not have any of the characteristics associated with Bills of Lading, and apart from its acting as a temporary receipt for the goods, it can be ignored.

6.6 MARINE INSURANCE

As the name implies, this is a form of insurance which indemnifies the owner of the goods against loss or damage to them whilst onboard ship, either in port or during a voyage. The value of the cargo insured might have to be estimated and agreed between the owners and the insurers before the loss or damage can be demonstrated to the home market (and circumstances can therefore occur when the owner may in fact receive a greater or a less sum than the actual value of the loss or damage).

Marine insurance covers all perils of the sea: fire, war or attack by nations not officially at war, piracy, seizure, restraints, jettisoning, barratry or any other peril specifically named in the policy concerned. It does not cover depreciation due to normal wear, tear, storage or the efflux of time, for example, food going bad, or fruit over-ripening during a prolonged voyage. (*Note:* The definition of terms more rarely used is found in Appendix 4.) Marine insurance (like other engineering insurance policies) is based on *umberrimae fidei* (or disclosure of all information) and the indemnity may be avoided by the insurance company if the owner does not comply.

The party taking out the insurance on the goods will naturally be the one having responsibility (as owner at the time) for the goods, and will thus depend on the mode of transfer which the parties have chosen to adopt. Thus, with FOB, it will be the buyer who owns the goods as soon as they cross the ship's rail; with CIF, it is the seller or exporter. Those are the parties who own the goods, and the parties on which any loss or damage will fall.

Equivalent air-freight insurances are obtainable which cover the specific risks during movement between the two named airports in the air consignment note, and during loading and unloading of the aircraft. The carrier will already hold insurance for the goods while held in his warehouse pending flight.

Both forms of insurance can be obtained through the carrier's or may be had from the seller's normal insurance agents. They are considered quite separately from normal insurances, and are undertaken by insurance firms specializing in the special risks which affect goods during the respective methods of transport.

6.7 ECGD

The Export Credits Guarantee Department is a British government-sponsored insurance organization which is prepared to take on those risks inherent in exporting which the ordinary insurance market is not prepared to handle. We have already spoken of the numerous risks which beset exporters and might otherwise result in expensive losses on export contracts. Mostly they are situations outside the control of the exporting firm itself, but the extensive intelligence network which the ECGD has established during its lifetime regarding foreign countries and their financial characteristics (and about many of the trading firms themselves), enables ECGD to proffer advice which a prudent exporter should follow. Their object is, of course, to encourage British firms to take on export business in spite of its inherent risks by guaranteeing their financial position, and thereby enabling them to enter competitive enquiries without having to include high contingency sums. If they did this and the risks continuously gave them losses, they would refrain from undertaking export business at all, let alone at competitive figures.

The ECGD has to operate on commercial lines but with some limitations. It cannot budget for a loss (which would be regarded as a government subsidy on exports – universally acknowledged to be a bad thing with no maximum); nor is it allowed by the government to make excessive profits (required for its prudent reserve against a succession of large claims in one financial period). The ECGD is committed to remain solvent over a period, and abnormal conditions can only be met by having such standing reserve. Otherwise it may be regarded as a non-profit-making organization.

The ECGD can deal direct with banks, export merchants or individual firms. Normally it insures against the following risks:

- Insolvency of overseas buyers arising after the seller has entered a contract with them.
- A buyer's refusal to pay within 6 months of the due date, in respect of goods delivered by the seller and duly accepted by the buyer.
- Legal or governmental regulations (outside the control of the seller or the buyer) which prevent, restrict or control the transfer of payments due under the contract.
- Political events, economic difficulties (including exchange abnormalities) or administrative problems (all arising outside the UK – i.e. are not present in 'home' contracts) which prevent or delay transfer of payments due.
- Cancellations of contract or delivery (outside the control of the seller and the buyer) or any import licence or permit to export foreign currency as specified in the contract, or the like.
- War, unprovoked attacks, revolution, civil disturbance or Acts of God* happening outside the UK, when the cause of loss or damage to the goods is from a risk not commercially insurable under other normally held commercial policies. Thus piracy on the high seas, a risk insured under marine insurance, would be excluded here.

Insurance cover arranged with ECGD normally extends to the whole of an export contract – i.e. ECGD comes on-risk immediately the buyer signs the purchase contract and pays the sum due on signature. It is at this point that the exporter is entitled to start work and amass expenditure on the contract, and from this point he is in a loss situation if the buyer refuses to pay up. The ECGD will not normally cover pure repudiation of a contract as such, this being considered a normal commercial risk and not peculiar to exporting. An exporter must still have his normal commercial wits about him.

The ECGD estimate their risks by considering the information they have (and it is considerable!) on most overseas countries, and the

* See Section 5.9 on 'Force Majeure'.

financial reliability of the most prominent industrial concerns. They can always get more detailed information, in specialized circumstances, through their contracts with commercial attachés at British embassies overseas. They will usually require to check the form of a proposed contract, to make sure the seller is not leaving himself open to fraud or disclaimer, or accepting imprudent methods of payment or the like. They obviously want to make sure no loopholes have been left in any proposed arrangements. Even so, they may refuse to insure at all, especially in territories which they know are experiencing severe payment difficulties. In some cases, they may go part-way, by reducing the percentage guarantees they will give, or they may require the seller to include certain 'special conditions' into a contract with certain territories, for example, a requirement for the buyer to obtain prior authority from his government to allow payment for the goods from his central national bank in sterling.

Alternatively, they may insist on the exporter obtaining prior guarantees of payment (such as by irrevocable Letters of Credit) at (or shortly after) the due dates, or the specification of exact dates in place of rather more unspecific periods. More usually, however, they will reflect the risks they are guaranteeing against or are likely to incur by the rate of the premium they quote in any particular case, which must be recovered by the exporter in his quoted price to the buyer. The ECGD will usually give especially favourable rates to exporters who frequently call on their services, or who automatically call for insurance cover for all contracts outside the UK, quite regardless of the identity of the buyer. In such a case, they are able to spread their risks (to the benefit of the exporter), those buyers with virtually no risk easing the situation with others having greater risks. The quoted premium is naturally quite high if an exporter only insures himself in the rare cases in which he realizes his danger is high. The ECGD also realize it and charge accordingly!

With this comprehensive cover and the large number of contracts to be covered, the system of insurance is fairly automatic, always provided certain stated requirements are met. Major contracts, that is above a certain contract value, or those with extended credit terms or any with exceptional (or unique) characteristics, are always considered on their own merits and the terms of suitable insurance cover arrived at by separate negotiation. In some cases, a loan is made to the seller by a bank, and the bank rather than the firm itself is the organization which needs to have its loan given extra security. Most other exporting countries, of course, have similar internal arrangements.

6.8 THE WORLD BANK

Although it is most unlikely that a project manager will ever be brought face to face with higher finance (in which we also include the World Bank), it is important that he should be aware of its operations and its methods. Officially it has the title of the International Bank for Reconstruction and Development (the Bank) and the International Development Association (IDA). Basically these joint associations comprise large financial interests prepared to make the necessary loans and credits to borrowers who otherwise would not be able to raise the substantial capital sums needed to undertake the very large national projects they aim at. They are especially active in impoverished or developing countries where they seek to improve development and local export trade. It is the old problem of 'We'd have more exports and earn more money if we had some money to produce more exports', and this is otherwise a vicious circle. Thus the World Bank and IDA not only provide a start, but they try to do so in such a way that it benefits their member countries, as well as the borrower's country.

By its Articles of Agreement, the Bank is required to make sure its loans are used for the express purpose for which they have been made, and the money is used with economy and efficiency. The ultimate responsibility rests with the borrower, who is (in general) required to obtain his goods and works by international contracts open to competitive tender by any firms in any of the member countries or Switzerland. It is not tied to such a procedure, however, and may substitute alternative arrangements with its borrowers when it is convinced that competitive tendering is not the most economic or effective solution.

Competitive bidding is first introduced by sending an enquiry to possible firms and by a policy of advertising in member countries with the object of discovering those firms which are both capable and interested in the proposed project. It is one of the many forms of pre-qualification bids in double-stage tendering. The borrower looks to see if the tenderer has these qualifications: (a) previous experience, (b) staff, plant and equipment to do the job, (c) organizational and administrative facilities for the project and (d) financial status. All firms judged to be suitable are then sent the full enquiry and are invited to put forward their individual tenders, including their proposals for financing their scheme, and the terms of payment they will expect. The World Bank requires to see, to modify as necessary, to institute negotiations and, finally, to approve the borrower's activities regarding:

- the original enquiry and advertising plans;
- the proposed tender lists;

- the draft contract documents; 'cost-plus' contracts are not, in general, supported as there is nothing 'solid' in their final aims;
- tender appraisal and negotiations by borrower;
- terms of contract eventually decided on.

Further details are given in their booklet, *Guidelines for Procurement under World Bank Loans and IDA Credits*, from their head office in Washington, DC, or their European office in Paris. (More details are given in our second volume, *Competitive Tendering*.)

The World Bank does not itself publish Standard (or Model) Conditions of Contract, as it would be difficult to propose one which adequately covered the range of projects they expect to foster and also maintain an international flavour. The choice is left to the borrower and his client and is subject, as usual, to the World Bank's approval.

6.9 PAYMENT OF THE CONTRACT PRICE

The matter of settlement of the contract price in a way which satisfies an exporter, and in a form which he can use to pay off all the expenses he has incurred, is just as big and important a point (bigger from his point of view!) as getting the goods to the buyer. One of the first things to be settled is the currency in which the contract price is to be paid, and where the money is to be made available to the exporter. The question, 'What currency?', may well be a matter of what currency the overseas country has available; it is not merely a matter of what appeals to the buyer, but what his country can give him. From the seller's point of view too, it is not just a matter of what he would like, but what his country (i.e. the Bank of England) can readily convert into what he might regard as 'spending money'. There are some currencies which are regarded as of little value to the country: 'we have piles of them already and very little counter-trade on which to use them'. In other words, the rate of exchange is low here but may be high in the territory concerned. Competitive prices in the UK will be non-competitive in the territory. In other words, do not choose that currency, but preferably one which we in this country want. Ideally, of course, we should prefer sterling itself, which is immediately transferable to meet our industrial needs, but equally we might be happy with US dollars, German Deutschmarks or even Swiss francs, all currencies for which the country has a big requirement and all of which have a reasonable stability of exchange rates. It takes two parties to make a contract, so agreement must first be reached on what suits both parties.

Most of the exporter's expenses in connection with the goods are incurred in England; payment should therefore be made in London. The other party might equally well opt for his capital city, but usually both agree that it is inadvisable to incur the extra exchange and local expense of paying over in some other foreign city. If payment is in sterling, then London is the natural choice of venue; the buyer then takes the risk of exchange fluctuations from his own currency into sterling. Sterling is today one of the world's most stable currencies and few buyers can find convincing arguments against its use. We must avoid the trap once run into by an unsophisticated Third World country, which (on the argument of relative stability) accepted a contract price in gold such-and-such. This was, of course, equivalent to being paid in a given weight of *gold*, and the price of gold was steadily rising at the time! The buyer paid dearly for his purchases. It is many years since the majority of countries left the so-called 'gold standard' and substituted a more stable relationship between the major currencies of the commercial world. For example, many currencies were tied to a fixed number of US dollars. In one or two cases, a local currency is maintained at a purely artificial rate by closing the frontiers to its export; an overseas trader can easily accept payment in the local currency and then not be allowed to get his balances out of the country. A large proportion of his working capital can easily be tied up in this way.

In the above, we have assumed the project manager and the seller of goods are both resident in the UK; in large projects overseas this may well not be so. The seller and the Employer (and his site), and his project manager, may all be from different countries; the problem of payments is the same – but more complex!

The contract price is usually paid by a series of instalments, usually tied to identifiable stages of the work. The following are typical for contracts overseas:

- A down-payment on signing the contract as a sign of firm intent. It is a great help to a seller of goods who might have cash-flow difficulties if his date of delivery is late in the programme.
- A percentage of the contract price against evidence that goods have been shipped, or handed to a recognized carrier.
- A percentage against completed approved drawings (if applicable to the particular contract).
- Payments against stages of erection on-site; usually on completion of parts of the contract works.
- Payments at time of take-over or occupation.

- Payments against successful completion, and prescribed tests on completion; a large project might well be taken over and completed in sections, making several phased payments under this heading.
- Further instalments payable on fixed dates (or after the efflux of fixed periods) following completion; these are the stages of a period of credit arranged by the Employer at the time the project is undertaken; his credit might, for example, be over 5 years following completion, after the project is put to work and earning money.

The 'down-payment' can be made by any normal means. The seller will have made no commitments prior to getting his contract, so he is not out-of-pocket. The Employer is often on hand to sign the contract documents and can have with him a sight-draft, a banker's order or a certified cheque. An uncertified cheque is not cash until it has been accepted by the Employer's bank, and this might take some time overseas. However, the Employer has known for some time that a source of ready cash was to be made available on signature, so there is usually no difficulty with the first (or 'down') payment.

The second is a more crucial one, as it is at this time that the seller lets go his security (i.e. the goods he has sold). We have already seen how he can put a withholding order on the shipment of a ship's cargo by suitably filling in the 'consignee box' of the Bill of Lading, but this solution is not open to the seller in road (TIR) deliveries or air-freight. He must either sell on credit or have the money in his hands in advance, or deal with the Employer's agent, exchanging his shipping documents for cash at the time of forwarding the goods. The agent holds the money until he has the carrier's receipt for the goods as on the invoice accompanying them. The corresponding extract from the payment terms of the overseas contract would read (for example) as follows:

X% of the contract price shall be due at the time of signing and shall be paid by the Employer in Sterling to the order of the seller at . . . bank in London within three days of becoming due.
Y% of the contract price shall become due against delivery of the goods as evidenced by clean bills of lading, air-consignment notes, or carrier's warehouse receipts (as applicable in any case) and . . .
be paid in sterling in London by means of an Irrevocable Letter of Credit to be established with . . . bank in London as soon as the seller shall have given the buyer 30-days' notice of his readiness to despatch, the Letter of Credit remaining valid for a period of 60 days from its date of drawing.

This example introduces a Letter of Credit.

6.10 LETTERS OF CREDIT

A Letter of Credit is just what its name suggests – it is a written letter from an overseas bank to a bank in the UK stating that the buyer has held available a specified sum of money to pay to the seller as soon as the latter satisfies the terms of the credit. It involves a series of banks, and without some explanation, these are apt to become confusing:

In London
1. The seller's normal bank
2. A special bank (or department of the seller's bank) for dealing with international transactions.

Overseas
3. A central bank appointed as a branch or a 'correspondent' for that territory by the international UK bank.
4. The buyer's normal bank.

The buyer deals with (4) which is in touch with (3). In its turn, (3) deals with the international bank in London (2). In effect, (3) becomes a branch in the overseas territory of (2), who accept its statements without further checking or delay. The seller normally deals with (1), but may for the purposes of the financing and documentation of this transaction be put into direct touch with (2). His normal branch (1) may not be set up to handle shipping documents, etc. itself and would anyway pass everything on to (2) for action.

This is the mechanism of the Letter of Credit. The buyer makes his arrangements with his bank (4), who establish their credit with bank (3). Bank (3) informs its principal (for whom it has been appointed 'accredited correspondent' that it holds such a credit, and they, in turn, notify the seller direct or his bank (1). The seller performs whatever is necessary to redeem the credit, and produces documentation to bank (2) to prove it. They then pay him his cash, and tell their correspondent that the credit has been expended. Bank (3) charges bank (4), who debit the buyer with a firm debit. The Credit is, of course, ended.

There are several types of Letter of Credit. A Sight Letter of Credit is one paid out in cash 'on sight' – i.e. without further action by the recipient. Thus traveller's cheques, banker's sight drafts and certified cheques are all forms of 'sight' documents.

An Irrevocable Letter of Credit offers the recipient (i.e. the seller) more security, as once the credit has been opened, its terms (and it itself) cannot be modified or cancelled without the written agreement of both buyer and seller. It is always payable as cash, as soon as the seller fulfils its terms.

Going a stage further in the degree of security is a 'confirmed irrevocable credit'. This is one in which the named bank in London adds its confirmation that (as long as the terms are fully and strictly adhered to) they will themselves make payments due under the credit, regardless of any upheavals, political, legal or economic, which might hinder payment to them from their overseas correspondent bank in the territory. All credits cost the buyer a premium; an Irrevocable Letter of Credit costs him more, depending on how long it is valid. Such a period should therefore be as short as reasonable, bearing in mind that communications with the overseas country might be extended. Post delivery may take even longer than in the UK. This is why, in our example above, the buyer was given 30 days' notice of our readiness to dispatch the goods to him, and he then had time to establish his credit in London before the contractually due date. If he were left to his own devices, he must include for delays in manufacture, in testing, in repairs or replacements as a result of the works tests, and so on. His Letter of Credit might need a considerably longer period of validity.

Confirmation in London is an additional service to the seller, who has to pay for it. No doubt, while the customer pays in the long run, he must not fail to allow for the charges in his quotation for the job.

Here we have followed just one chain of banks; there is an alternative chain which may be met with and which affects the same situation in practice. The buyer's bank (4) may have its own 'correspondent' bank in England through which it deals. This correspondent bank communicates with (1) direct, and handles the export documentation, etc. on behalf of the seller and the buyer, in fact the London end of the transaction. In effect, it is the same, except that the overseas bank deals through its own correspondent in London instead of through the London bank's correspondent in the overseas territory.

6.11 BILLS OF EXCHANGE

A Bill of Exchange is a 'promise to pay'; it does not itself represent a cash transfer, but has to be backed up by one when it matures. It has been defined as (Bills of Exchange Act 1882):

> An unconditional order in writing, addressed by one person to another, signed by the person giving it, requiring the person receiving it to pay on demand or at a fixed or determinable future time a sum certain in money to, or to the order of, a specified person, or to bearer.

It is therefore drawn up by the exporter (the seller), who is referred to as the 'drawer of the bill'. It instructs the buyer (importer) to pay a specific sum, in a specified currency, to a payee at a specified place on demand, or at a stated (or determinable) future date. In some cases, the payee is the drawer also; or money could be paid to the order of the drawer – e.g. to his nominated bank, the bank becoming the actual payee. Note that a Bill of Exchange must be 'unconditional' and cannot express any stipulation (or it is not a Bill of Exchange), but may relate to the condition of the goods before it can mature. The sum to be paid is also to be 'a sum certain' and cannot be an approximation, or have unspecified additions – e.g. 'X pounds plus bank charges' would be inadmissible.

If payment is to be made 'on demand', this is known as 'on sight' and the bill becomes a 'sight draft'. When payment is to be made at a future date, it is called a 'tenor bill' and the forward period to maturity is the 'tenor' of the bill.

The bill, drawn by the seller, is given to the buyer in the overseas country, who 'accepts' it by signing across the front of the bill and dating it. This is the 'sight date' and any fixed period to maturity (e.g. a 60-day bill) will date from such 'sight date' or 'acceptance date'. By signing the bill in this way, the buyer:

- acknowledges that he owes the drawer the sum of money specified in the bill;
- agrees to pay the drawer (or to his order) the stated sum of his debt, at the time and place due;
- agrees to obtain the currency specified at his expense and to pay the bill in that currency;
- gives the drawer written proof of his indebtedness on which the drawer can sue if the bill is later dishonoured.

Note that the buyer is paying no money, but in much the same way as he would with a forward-dated cheque, he goes through the motions of paying his debt, but he can still fail to keep his promise. A dishonoured bill 'bounces' like a dishonoured cheque! No money at the appointed date, and there is an unpaid debt which can be pursued by taking the buyer to law (at a hideous cost). A Bill of Exchange can still be dishonoured after it has been 'accepted', though such a bill is considered to be the highest form of commercial undertaking. A trader who dishonours an accepted Bill of Exchange would not remain a trader for long!

The 'due date' specified in the bill is by custom given a period of grace of three days, or rather three *working* days. Once a bill has become a firm promise (i.e. it has been 'accepted' by the importer (buyer)) it is a negotiable document. That is, it can be sold time without number, and cash obtained by the person disposing of it. If the seller wishes to make use of this facility, he cannot expect to receive the full face-value of the bill itself, but a reduced amount. In the first place, the bill only becomes worth the full amount at its date of eventual maturity when the acceptor pays up the amount he has accepted. Prior to that, the bill has a 'discounted' value (which we have dealt with elsewhere – see the second volume, *Competitive Tendering* in this series). The purchaser may also deduct any 'risk contingency' he considers advisable, and his own profit on the deal. If the purchaser is a bank, the whole transaction is equivalent to receiving a bank loan on the security of the Bill of Exchange, and the seller can expect to pay interest on the loan at the usual bank rates.

Bills of Exchange are frequently drawn in 'sets' of two or three, the number depending on the subsequent expected needs of the buyer and the security of his communications with the seller. A single copy might be lost or misplaced or delayed in transit, so that duplicate transmission would be resorted to in an attempt to avoid the numerous difficulties which could ensue. The importing country might need several copies to conform with its currency restrictions. Each copy of the set is identical, except for its references to the other copies of the set. Only one copy of a set is 'accepted' by signature of the buyer and redeemed by a cash payment at maturity. Otherwise the buyer might find himself paying the same bill twice!

A bill does not require any special document as long as all the necessary details are stated within it. Printed forms can be obtained from any good commercial stationers, but a firm's plain typing paper is equally acceptable. A typical Bill of Exchange might read as follows:

Bill of Exchange
(Copy 1 of 3)

Amount: £200,000 *No. of Bills:*.................... *Date:*....................

180 days after sight, pay this first of exchange (second and third of the same date and tenor being unpaid) to the order of our account at .... bank situated at .... in London the sum of two hundred thousand pounds sterling, the currency of the United Kingdom for value received and in accordance with the terms of payment agreed between us in our contract dated ....

All charges in connection with foreign currency exchange are to be at your account.

Addressed to:	*Accepted and signed:*..............................
(Name and address of)	*for and on behalf of:*...............................
(Importer–Buyer)	
	....
	Date:..

7 *Subcontracts, preferred subcontracts, nominated subcontracts, prime costs, contract assignment and insolvency*

7.1 SUBCONTRACTING, SUBLETTING

Subcontracting is an essential part of the way all contractors operate. None has the facilities or money to lay in stocks of the numerous every-day commodities it has to use in its routine jobs; these it gets as it requires them by subcontracting. At the other end of the system there are companies which are recognized by their employers as being nothing more than entrepreneurs rarely make anything themselves, but acting as coordinators' between their many subcontractors. They mostly erect plants having a family resemblance to one another, and by using the same subcontractors for similar parts of them, they can offer their clients a much quicker project turn-round. This is, perhaps, an illogical extension here, but they have converted a process of sub-contracting, into one of *subletting* the whole project works, with them-selves still acting as the main contractors and taking full responsibility for the works.

However, in this case as in many others, a surfeit becomes an up-setting thing in an average engineering contract. Nothing is held against the subcontracting of every-day commodities, the 'sticks and string' of constructional work or specialist equipment and expert installations; all Employers realize that these items must be sub-contracted by all normal general engineering firms, but the latter draw a line at employing subcontractors (outside the reach of the Employer) for work they could well do themselves under direct supervision. In an average engineering contract, an Employer has direct recourse only to his main contractor, and it is on him alone he can apply his immediate supervision and his pressure to keep to the programme. He is not content to sit back and make sure his main contractor will bring all his subcontracts together at the right place and time.

The Employer is always conscious of the fact that any subcontractor which he has not himself 'vetted' may renege or become insolvent.

Although this may still be a matter for his main contractor (who has the subcontract), it is his own project which will suffer in the long run by the delay which ensues before another is appointed.

It is, however, a matter on which an Employer must make his wishes clear to his tenderers at the time he sends out his enquiry. This will usually state or imply:

- The chosen contractor will only subcontract a *part* of the contract works, with the prior permission of the Employer or project manager in writing. Such permission will not be unreasonably withheld.
    - (a) This requirement is not necessary in the case of materials, etc. in general supply and which it is usual to subcontract in the industry;
    - (b) Subcontractors envisaged at the time of tendering to be given in a pro-forma attached to the tender and forming part of the offer.
    - (c) Subcontractors proposed after the signing of the contract (e.g. to deal with fresh work or variations) will require separate permission in writing in each case, coming into line with (b).
- The contractor will include in each of his subcontracts a clause permitting the Employer to visit a subcontractor at his place of work in order to approve the procedures he adopts or to discover the progress made with the work and his anticipated date of delivery to the main contractor.
- The contractor will not sublet his contract or any material portion thereof, except as previously approved.
- The contractor's liabilities and responsibilities towards the Employer shall not be affected in any way by the subcontracting he or his project manager may approve. The contractor shall remain responsible to the Employer for all acts, defaults or neglect occasioned by his subcontractors.

The above restrictions apply chiefly to constructional or building work in which the Employer provides or approves the fully detailed plans for executing the works. They are not usually applied rigidly to contracts for the supply of plant or machinery, where the Employer confines his specification to functional requirements and any necessary ruling dimensions. The contractor provides his own details of actual construction. In such cases, the contractor is carefully chosen and his normal factory procedures (including those for subcontracting, treatment or assembly of components of the contract works) are to be applied in the Employer's contract. In other words, he is left to 'get on with it his own way', using his normal factory methods.

In most cases, the subcontractor's agreement to permit access to the Employer's representative is as much as an Employer can expect if he is to retain the main contractor in his role of responsible contractor: in exceptional circumstances, the Employer might obtain consent from the contractor (and get the subcontractor to agree) to his signing a collateral agreement direct with the subcontractor by which he has control of the design or delivery date for the subcontracted works. We shall be saying more about this in a later section below. The Employer can only continue to hold his main contractor responsible if he himself is unable to interfere between them (except when invited to do so), and any collateral contract must be very carefully worded to avoid any danger of his accepting direct responsibility himself.

Whenever a pro-forma return of proposed subcontractors is included by the Employer in his enquiry documents and has to be filled in and returned by the tenderer for approval, the pro-forma is an integral part of the offer, and later acceptance by the Employer must imply that the subcontracting shown on the form, both as to extent and as to the nature of the parts subcontracted, has been approved. The main contractor is therefore bound by the contents of the form, and is not at liberty to add to or change them without separate approval from the Employer or his project manager.

In his consideration of the proposed subcontracting by the main contractor, it is important that a project manager has some yardstick by which he can compare proposals from one source with the others. The following are the more important aspects on which a project manager must satisfy himself before signifying his approval of any subcontracting proposed:

- Is the subject of the subcontract one he would expect a good contractor to carry out himself? Has the work been 'farmed out' because it is considered to have a low priority by the contractor? What were his reasons for not undertaking it himself? Are there any advantages to the Employer in the work being done by the proposed subcontractor?
- Is the proposed subcontractor technically suitable (a) to understand the work allotted to him, (b) to make a good job and (c) to get it to work on-site unaided?
- Is the subcontractor's financial stability acceptable, his claims reputation good and his labour relations sound?
- Are there any commercial, political or security implications in the use of the proposed subcontractor? Does he have any relations with the competition, and could he be a 'spy in the camp' if employed?

- Has the Employer any close relations with an equally suitable subcontractor. What are the pros and cons?
- Are there any subcontractors in the list of preferred subcontractors with better potentials, such as locations, reputation, experience, prices or precision or specialized work called for by the nature of the present duties (e.g. work under water)? If a *new* name is suggested at random, there is the risk that the contractor would treat it as a *nominated* subcontractor, with all that that implies.

A project manager must always weigh up whether it is worth the results of interfering with a potential main contractor in order to rectify the objectionable point. Personal approaches may suffice. It is accepted that many of the above criteria are, strictly speaking, the affair of the main contractor. To put it more bluntly, the contracts are 'his affair' and exist between him and his subcontractors. Nevertheless, it is the Employer who will suffer in the long run: if a subcontract is late, it is the Employer's project which will reflect the lateness. If the workmanship is bad, it might be good enough to pass the immediate site tests, but it is the Employer who will suffer the early breakdowns of the future. In his watchfulness for the Employer's interest, it behoves the project manager to keep a careful eye on the contractor, who might be aiming to cut his costs by seeking help from cheap firms outside. Not every contractor is 'bent' in this way, but it only takes one of them to spoil a project – and this must not be the one.

7.2 PREFERRED SUBCONTRACTORS

An Employer may include in his enquiry a list of his 'preferred subcontractors' for certain portions of the works he is proposing to let to his main contractor, or for the supply of certain items of materials. In this way, he hopes to avoid the need for using a nominated subcontractor. Of course, with enquiries for the supply of *plant or equipment*, organizations such as 'nominated subcontractors' are not used by manufacturers. The contractor puts forward his best suggestion to meet the functional specification included in the enquiry. All the same, the Employer might wish his plant contractor to use certain makes of accessories, and in such an event he can still use the 'preferred subcontractor' device.

The reasons for using 'preferred subcontractors' instead of allowing his tenderers a free choice may include one or more of the following:

- To obtain products similar to those already used at the site, thereby reducing the need for further spare parts and different maintenance schedules (e.g. instrumentation, small motors, pumps, etc.).

- To employ organizations he has used with satisfaction on previous occasions, or indeed is already using at the site on other work.
- To use other subsidiaries of his industry-group, or firms with whom he does a substantial counter-business. There may be advantages when dealing with countries abroad to institute a counter-trade and avoid losses in exchange transactions.
- To employ firms known to be already active in the site locality, especially in remote places overseas.
- To continue to use a firm with whom he has carried out development or research work, and make all possible use of existing liaison when his project designs involve or are built around his earlier co-operation. His main project contract gives him no other contact with his earlier work.

A tenderer can assume when receiving an enquiry containing a list of preferred subcontractors, that each nomination has been separately approached by the Employer and has signified his willingness to undertake the requisite work. From the list a tenderer is expected to select the firm he chooses, or which might give him the greatest discounts. He can liaise with the Employer if he thinks it advisable to do so before making his final selection. He is not *obliged* to select any one of the Employer's preferred list and can put forward his own nominee, but the Employer will expect him to have a good reason for so doing, and indeed may mark the offer badly if he is not given a reasonable excuse. The Employer has made his preferences clear: the tenderers would be prudent to heed them!

7.3 NOMINATED SUBCONTRACTORS

These have been labelled 'ugly creations' by a well-known barrister with wide experience in constructional contracts. The advice has also been given (largely because of the unsatisfactory contractual position), 'when in doubt, don't'. From which it will be readily appreciated that while they are quite extensively used in the building and constructional industry, they are not popular among contractual and legal individuals.

Historically contractors have long needed to use specialized subcontractors, and along the way the Employer had the bright idea of choosing such subcontractors himself, especially if he had to make some prior technical developments with them before putting them to work on his project. From this conception the Nominated Subcontractor was born. In a way, it was a question of the Employer taking for himself the best of both worlds – he wanted his main contractor to be completely responsible to him for his project and do all the bookwork (leaving him free), and at the same time, he wanted to dictate to his free-acting main contractor just whom he should employ as one of his subcontractors. Clearly, the main contractor needed some form of safeguards in such an arrangement, and so the idea was limited to the area which the Employer understood and which he was to design in detail himself; that is, the building and the constructional activities, and the safeguards were taken care of by special clauses in the conditions of his contracts. With factory-produced plant and machinery, the Employer was a bit more out of touch with what was practical in the manufacturing process, and hence did not interfere so completely; he was content in most cases to go as far as 'preferred subcontractors', with no compulsion on the manufacturers to make use of them if they were not readily able to introduce them instead of their normal suppliers. Standard conditions of contract for plant and factory-produced equipment therefore are rarely found to contain references to 'nominated subcontractors' or to any safeguards if they are ever to be used.

An Employer may choose his own subcontractor for one of a number of different reasons:

- To choose the most competent firm, either technically or on price, within his own experience.
- To employ a sister-company, another subsidiary of his own group of companies.
- To permit an earlier start on long-delivery works, ahead of appointing his main contractor. He starts a contract with the subcontractor to enable work and expenditure to proceed; his main contractor inherits it.
- To carry out joint development work with a firm and then ensure it is used for the resulting scheme when his project is executed.
- To standardize on a product already used on-site. Both spare parts and maintenance are aided thereby.
- To assist his main contractor by nominating a firm available in the site locality, and known to be reliable.

The firm is eventually 'taken on' by the main contractor as his 'nominated subcontractor' (always assuming the said firm fulfils the provisions made in the main contract to safeguard him), and mutual terms similar to those previously agreed with the Employer are accepted by them both. The main contractor is recompensed by a so-called prime-cost sum of money (see Section 7.4, below) in his bills of quantities.

The arrangement can work satisfactorily if, and only if, there is a certain amount of mutual trust:

- The main contractor finds the nominated firm to be acceptable (and vice versa), and he can negotiate a satisfactory subcontract with him.
- The subcontractor is prepared to accept the main contractor's conditions of contract (back-to-back), and that prices and payments agreed earlier with the Employer continue to be effective in the new conditions, or approximately so.
- The subcontract work, as approved and designed in earlier work with the Employer, still works satisfactorily after erection on-site under the nominated subcontract.
- The subcontractor is not in delay in producing his works by the specified completion date, fixed by the main contractor to meet his own requirements.
- If, for any reason, a nominated subcontractor fails, the main contractor expects the Employer to nominate a replacement and to accept liability for any hiatus or delay that is caused to the project or to the main contractor's operations.

Once a main contractor has issued his subcontract, the former Employer has no further direct contract with the activities of the subcontractor. Just as with any other project subcontractor, responsibility lies with the main contractor, and there is no direct link with the Employer through his main contract. This is in spite of the fact that the main contractor has taken the subcontractor on against his better judgement or 'under sufferance' on the Employer's recommendation and instructions. Not even the latter can *make* the two parties agree to accept any pre-imposed terms – or for that matter, each other! The whole matter has to be thrashed out, then, on an informal discussion beforehand as to what is and is not acceptable to both of them. They must be allowed to negotiate together to some extent, but not too freely. There have been numerous legal actions which, far from clearing up the position, have introduced even more extensive problems. As one barrister of long experience in the field has said, 'It will need careful consideration whether nominated sub-contracting can continue to be used with prudence having regard to recent court decisions.' Some indication of the difficulties which may be encountered can be indicated by the scope of the applicable clauses of the popular Institution of Civil Engineers (ICE) Standard Conditions (5th edn), clauses 58, 59A, 59B and 59C, and, in particular, to their effects on responsibility for design (especially if the Employer's design fails), contract delay, amounts of liquidated damages, performance tests on subcontracted work, point of take-over by the main contractor, and such like. The position is even more aggravated when the main contract (being a constructional one and using standard conditions appropriate to such work) is forced to use a subcontractor nominated to provide accompanying plant, which he has always done hitherto on a suitable set of plant conditions. This can happen frequently, and the plant manufacturer says, 'You *want* to use my plant, all right, you accept *my* terms and conditions'. Typical examples might be an automatic electronic-controlled traffic-signalling system as part of a roadworks construction project, or an automatic computer-controlled elevator bank to be installed in a new building block (being erected under a main contract using building conditions). The basic conditions of building works and plant works are so different that a back-to-back arrangement is quite impossible from the start, unless one or the other side compromises.

The most common 'sticking points' are the amount of Liquidated Damages (those for the main contract often exceeding the whole contract price of the subcontract), and responsibility for technical performance. The main contract might well use the ICE 5th edn, but the nominated subcontractor might equally well stick out for a BEAMA electronic contract, or even one of his own design which long experience has shown necessary for his specialized work. The main contractor might accept no design responsibility for the nominated subcontract. The misfits between nominated subcontractors and the main contractor are clearly shown up, leaving much to be covered by mutual trust and co-operation rather than by legal precision. The situation can be improved to some extent by introducing a direct agreement between the Employer and the nominated subcontractor, or some such arrangements as follow, always provided that the people concerned are amenable to legal co-operation in the ways suggested: it takes two to make an agreement!

- The Employer, early on, arranges a Bond with the firm he proposes to nominate, who promises to maintain his prices and deliveries, etc. in his subcontract when the main contractor is eventually chosen. But at the moment he has no idea who it might be! (See Appendix 11.)
- The Employer enters a collateral contract direct with the subcontractor, entitling himself to enter and take charge of factors affecting delay, change of design and performance responsibilities. Of course, this weakens the powers of the main contractor for the project as a whole and would need his active connivance.

- Careful and explicit wording of any development contracts between the Employer and the firm to be eventually nominated. Of course, at that point in time, the results of the collaboration might be vague and uncertain.
- Extensive professional alterations of the standard subcontract used by the main contractor with his non-nominated subcontractors to make it fit the reasonable complaints of a nominated *plant*-maker.

The above may all seem rather artificial, but they are practical efforts to bring two quite dissimilar situations together in a quasi-legal way. In some cases, the act of nomination by the Employer may be sufficient 'consideration' to make their contract valid, but in others it might become necessary for sums of money to be paid to produce the necessary agreed consideration. We shall be dealing with the questions raised in greater depth when we deal with Competitive Tendering in the second volume of this series, but the commercial relationship between the parties is still obscure. The only solution appears to be: 'if in doubt, don't'! Most of the time preferred subcontractors offer a much tidier solution to the problem.

Insolvency is a situation which cannot be ignored, even when the Employer has had earlier dealings with the nominated subcontractor; such a matter introduces special responsibilities, and these are set out in Section 7.6, below. If for any reason a contract between a nominated subcontractor and the main contractor fails, the latter expects the Employer to make a replacement nomination, and also to accept responsibility for any delay or extra cost resulting, both in the project and in the main contractor's operations.

7.4 PRIME COSTS

These have, *per se*, no legal meaning, and must therefore be defined afresh in each contract. They are the estimated cost of a nominated subcontract as made by the Employer and are paid to the main contractor against a separate series of three entries for each, in his bills of quantities for the main contract. Taken together, they form the total prime cost of a single nominated subcontract to the Employer. The three items are:

1. the price actually paid to the subcontractor by the main contractor for the works;
2. the price charged by the main contractor for his efforts in running the subcontractor and any site help supplied;

3. the main contractor's profit margin on the nominated subcontract itself, plus any other out-of-pocket expenses he might incur.

The question of any discounts allowed by the subcontractor is settled by the main contractor deducting them from the subcontractor's invoices and crediting them to the Employer, the one exception being any discount allowed for prompt payment of the account. This is earned by the main contractor making his payments promptly, and he is entitled to retain it himself. Any difference between the amounts actually concerned and those estimated by the Employer when the bills are drawn up are taken care of by the issue of variation orders, so that the two sets of sums agree. The second two items shown are usually expressed as a percentage added to the sum shown in the first, and hence will both vary in sympathy with it. Any requirements in the conditions of contract for contract price adjustment (CPA) can also be made to prime cost sums by the same procedure. The total prime cost sum is an essential part of the contract as a whole and cannot be omitted at the discretion of the Employer or his staff: the contractor is entitled to be paid it (subject to any modifications introduced by agreed variation orders as indicated above), and it cannot be removed.

7.4.1 Provisional sums

These must be carefully distinguished from prime costs. As the name suggests, provisional sums are moneys included in the project in respect of various features which it has not been decided to adopt at the time the contract is placed, but which later might be found to be advisable. They will then be given to the same contractor to carry out. They are included at the early stage for two main reasons: first, to get the contractor's agreement at the same time that he signs the contract, so that if it is later decided to have the features, he will be prepared to carry them out on his existing contract, and an approximation of the additional price. The second reason is for the Employer to get an estimate of the overall cost of a likely project, and to include the full amount in such financial arrangements as he may make. It saves having to go back with new requests later, which might cause both embarrassment to the financial participants and a delay in the carrying out of the project itself. With financial arrangements including all provisional sums, the way is clear to go straight ahead without incurring any delay in getting financial approval.

From the point of view of the project manager, he is free to use provisional sums or to omit the features concerned at will. The contractor has no right of expectation, and no assurance that such sums form part of his contract price. Provisional sums are not put into the bills of quantities *en bloc*, but are allotted each one to its own individual feature.

7.5 CONTRACT ASSIGNMENT

Assignment is the transfer of a benefit, to which a party to a contract is entitled, to another person (third party to the contract). A responsibility or a liability owed by a party under his contract cannot be assigned or transferred, the bargain of a contract cannot be transferred unilaterally.

The effect of the assignment of some benefit to which the party becomes entitled under his contract is that the third party (the transferee) can now sue directly the first party to the contract for payment or transferrence of the benefit. That is, an Employer may be sued under an engineering contract for non-payment of the contract price (or some part of it) whenever it should become due. He can no longer hold in abeyance the actual payment of an amount still in dispute, if for example, his contractor is being difficult about the completion of a stage of the work. An assignee can step in and demand payment in full on the contractual date. The following points must be noted in connection with an assignment:

- There must be an intention to assign: coercion or demands by the 'assignee' are not permitted.
- Assignment is absolute: no conditions or 'strings' are to be attached. A contractor divests himself of his entire interest in the benefit assigned.
- An assignment of benefits arising from a contract can be made by the party receiving them without reference to the other party with whom he is contracted. That is, a contractor could assign his benefits under a contract without reference to the Employer.
- If an assignment is by contract (and not by 'gift'), the contract needs all the properties of any normal contract – e.g. there must be a two-way consideration.
- An Employer can insist on a tenderer accepting a term which prohibits assignment as a prerequisite to awarding him a contract.

Most engineering contracts include a clause prohibiting subletting and assignment. What a contractor does with the contract money after he gets it is no matter for the Employer, to whom any third party to whom it might be owed does not legally exist. He is naturally interested to know his contractor's financial weakness right up to the time his contract is completed, but there is not much he can do about it at this stage. He might, if he likes to take the risk, advance his contract payments, with an agreement that such advances are used solely for his own works, but this does not by any means remove all the chances of insolvency.

The 'prohibition clause' might appear a little strange in the contract conditions. A contractor is assured of his payments to do with as he likes, and it has long been recognized that he must to some extent subcontract (see Section 7.1). His contract already prevents him subletting or subcontracting without the prior approval of the Employer, and this must be obtained for each occasion subsequent to the list of subcontractors he submits with his tender. The clause must therefore be interpreted in a narrower way, as preventing him getting out of his basic responsibilities by shelving them on someone else, and he must not give the Employer the additional difficulty of dealing with some unpaid third party. A project manager should note that, without the agreement by the prohibiting clause, there might always be an assignee (of whose presence he has been unaware) dunning him for direct payment, on time, of the contract payments. He should also be aware that while an employer cannot assign an employee's contract of employment, the employee can always assign his wages. However, he cannot do so if he is a *public official*!

7.5.1 So-called 'assignment' of complete contracts

Legally an assignment can be made only of benefits. Responsibilities cannot be assigned, and as we have already said above, a contractor is saddled with the responsibilities he has undertaken under the contract, even if he is allowed to subcontract the actual work by the Employer. He still remains responsible.

If a party to a contract is 'taken over' and ceases to trade, the Employer is faced with a decision as to what he can do. Usually there are three possibilities:

1. If the new owner is ready to do so, he may transfer the responsibilities to the new owner and continue with the contract-works as before. The new owner then takes over the contract when he takes over the contractor, but the Employer is presented with a 'strange' contractor.
2. The Employer may determine his existing contract and make a new one with the new owner. He may not get such good 'terms' or he may get better ones!

3. He may enforce the existing contract and sue for non-performance. He can then sign a new contract with anyone he pleases. But can the old contractor find the damages in cash?

The Employer will doubtless be swayed in his decision by the circumstances and economics of the three possibilities, but in most cases possibility 1. will be the most attractive. It will save most on delay and the new owner will probably be a firm of substance. Technical ability might be the most doubtful consideration. 'Specific Performance' is not to be otherwise expected (see Section 5.5), especially if delay is important.

7.6 INSOLVENCY OF CONTRACTORS

The Insolvency Act 1987 was expected to providc a concise and unified code of operation for matters of insolvency. In the event, the word 'concise' proves to be hardly the correct one as the Act runs to 444 sections, in dense legal language, plus 15 appendices, or a total of 321 A-4 pages. Its table of Contents is most useful in finding one's way around, but a detailed index is sorely needed! The Act deals with all companies, both limited and unlimited, and with partnerships and individuals; it covers voluntary liquidation, liquidation by the courts, the demands of creditors, receiverships for debenture-holders and other aspects, including the avoidance of fraud and deception.

The overall effect on engineering contracts is likely to be small, but so far it has not been severely tested. It does, however, bring into question immediately the wisdom of including in a contract (as do so many Standard Forms of Conditions) the automatic determination of a contract whatever the nature of the contractor or his type of insolvency. Such conditions should be given further detailed examination by legal experts before being accepted at face value by project managers.

There are a few fresh conceptions which need to be recognized, as follows.

(a) A new type of professional is introduced, the Insolvency Practitioner, appointed by the Secretary of State, either through selected professional bodies, or personally as 'fit and proper persons with the right practical training and experience'. In general, such professionals will always be appointed by the courts to fill 'management' positions with insolvent bodies.

(b) An Insolvency Practitioner's Tribunal is to be maintained by the Secretary of State to which complainants can appeal. Each tribunal will

contain several insolvency practitioners from a panel maintained by the Secretary of State.
(c) Any body suffering from financial difficulties has an opportunity of instituting a 'voluntary arrangement' or a scheme for its affairs under a 'nominated trustee', which (if approved by everyone concerned, including the courts) will be given a trial run to see if the body can be made viable. If not, proceedings for insolvency will be authorized by the courts, but meantime all such legal processes to this end will be suspended. It is given a clear run 'under new management'!
(d) When insolvency proceedings are begun, the court will replace the Trustee by its own insolvency practitioner, who will act as manager of the winding-up process and the liquidation. At the start, the Official Receiver for the court may himself step in, then hand over to an insolvency practitioner.
(e) Individuals declared bankrupt will have their affairs managed by a Trustee in Bankruptcy, but will normally be automatically discharged in 3 (or in some cases 2) years time.

In all cases, the appointed official will take over existing contracts, but he may disclaim 'onerous property' which can include unprofitable contracts. These he may replace by freshly negotiated ones, or he may choose to remain in breach of contract (if that seems to him to give better results for his creditors or for the company). The Employer is then entitled to claim his damages for such a breach of contract, but as he will only become another non-preferential creditor, his chances of recovering his losses and time on the project are indeed slim.

The position of a bankrupt individual is somewhat different. While he may not 'trade' on his own behalf, he must nevertheless hand over to the appointed Trustee anything not needed to support himself and his dependent family. Any contracts of a personal nature (e.g. a personal agency agreement, a contract of employment) are not transferred but automatically become null and void. He can – and usually does – go ahead with his employment in which he gets wages for working for another. If necessary, the Trustee in Bankruptcy will institute new contracts of employment and give him another job. He may still, as a bankrupt individual, face claims arising from his former employment (possibly even from his new manager).

The newly appointed liquidator can be expected to be quite ruthless in his search for suppressed assets, which he can use for the benefit of his creditors or his intended activities. A project manager may have considerable difficulty in establishing his right to any property, especially such objects as may be lying for one reason or another at the premises of the insolvent firm. Still less is he allowed to recover possession of them or try to establish a firmer claim as the holder of them. It must be rigid practice for project managers and their staffs (even if they are not faced with immediate insolvency proceedings) for all the Employer's property left outside their own control to be suitably marked, listed and recorded with the body presently holding the same. Verisimilitude is made more immediately clear if some identifying marks, such as serial numbers, are also recorded. A suitable wording for a firmly affixed label is as follows: 'This .... (identified by a mark) is the Property of .... [Employer]. It has not been deposited by way of collateral for any loan or favour by .... [the Contractor] on behalf of the Employer or anyone else. The Employer reserves the right to enter the premises of .... [the Contractor] and to seize, remove and recover the item named at any reasonable time of day or night, with or without prior notice of his intention having been given.'

Special attention should be given, in this way, to the prompt marking of any finished work (or partly finished work due to be supplied by a firm shortly to go bankrupt) which has been wholly or partly paid for, but not yet delivered. Attention is drawn to the definition of 'property' given in Section 1.9 in this book, and the need for a clear clause in the contract defining the exact conditions on which property moves to the purchaser (the Employer). Any goods or materials outside this definition belong to the contractor and the new official can rightly take possession of them to the dismay of the project manager. (See also Section 3.6 of the present volume and Section 4.19 of Volume 2.) Of course, if the project manager's staff can get pre-knowledge of any impending insolvency, any such property should be removed and appropriate payments made immediately, but this situation is unusual: appointed officials act with immediacy but without advance publicity. Cash problems by a contractor may be a good guide. Otherwise, an experienced project manager, using his engineer's nose can learn much from a 'visit' to a manufacturer's works, by 'off the record' conversations, observing the state of activity and bustle in the shops, the amount of work passing through, and so on.

7.6.1 Insolvent subcontractors

Subcontractors selected and appointed by the main contractor have no special characteristics. There is a valid engineering contract between them, and any responsibilities the main contractor may have to his Employer are dealt with quite separately. He has undertaken certain dates and duties, and will fall down on them if his subcontractor becomes insolvent, unless he takes firm and immediate remedial action. Just why the subcontractor became insolvent is no concern of the Employer: he wants to know what the effect on his project is going to be; he will eventually collect from his main contractor any damages or agreed liabilities which his contract prescribes and meanwhile urges his main contractor to do whatever he can to ameliorate the situation.

The situation regarding nominated subcontractors is quite different. Here the subcontractor was selected and appointed by the Employer, and 'wished on to' the contractor, sometimes even against his better judgement. The main contractor has no freedom of action in the matter and merely gets his instructions. The position in respect of breaches of his main contract with the Employer is a more complex one. The effect on the project of the insolvency of a nominated subcontractor is entirely one for the Employer himself. The main contractor 'marks time' and awaits the nomination of a suitable replacement. This, and any costs incurred in the change-over to the substitute, are charged to the Employer. The situation is made worse if the main contractor has elected to bear no design responsibility for the works provided by the nominated subcontractor. The whole business is one more example (if such were needed) of why nominated subcontracting should be avoided!

7.6.2 Contractor's materials and plant brought to site

The chances of the Employer getting any cash return to meet his extra expenses when his contractor becomes insolvent, are very slim. All the more important to him are his chances of taking possession of any materials or plant belonging to the contractor which have been previously brought to site and left there. Once built into the contract works (strictly speaking, fixed to the site), the materials all belong to him, whether he has paid for them or not. It is a basic element of property law that when something is 'attached' to the works, it automatically belongs to the landowner (in this case, the Employer). It is the loose items he is concerned about, contractor's plant and unused materials.

The Employer's contract is probably based on some edition of Standard or Model Conditions of Contract, and they will usually have a vesting clause making all the contractor's property on site automati cally his. Some conditions are more far-reaching than others (e.g. the ICE 5th edn, clause 53) and they all apply to things *owned by the contractor* (i.e. both for future incorporation into the works, and what he has brought to assist in the execution of the works but does not incorporate into them). By a legal quirk of the bankruptcy laws, the Employer will probably hold out against the official appointed to take charge of the insolvent contractor's business and be allowed to retain the goods as security for the loss he expects to sustain as a result of the contractor going insolvent. It is therefore important that he should prevent by all means the removal of these goods from his site, both by the liquidator and (earlier) by the contractor himself before the project manager or his site representative has become aware that the contractor is in financial difficulties and likely to become insolvent. A further difficulty is that the vesting arrangement does not apply to those goods not *owned* by the insolvent firm, for example, anything on hire or borrowed. These can be claimed and must be given back to their rightful owners.

However, the main doubt was introduced by the Court of Appeal which ruled that if a contract to buy and sell included a clause of a certain type, that clause was valid and meant that the buyer did not have title to the goods until he had fully paid for them (This type of clause is always referred to as the 'Romalpa' clause, one of the litigants in the case in question). So that if the contractor has obtained any of his materials or plant by a contract containing such a clause, and has not paid for them 100% then he does not *own* them, and the vesting clause no longer applies to such goods. The contractor and his official caretaker can remove them, and repossess them as their own. There have been a number of successful cases in which Romalpa clauses have been overcome, but the Court of Appeal ruling still casts a considerable doubt over the whole question of what is, and what is not, vested in the Employer. There is still confusion, what with goods on hire, hire-purchase, Romalpa clauses, and so on. The Employer still is entitled to any 'performance guarantee' he may have been given by a third party, and the bankruptcy court official may decide to continue with the contract. Indeed he may attempt to re-negotiate it on terms more favourable to himself than the former one (but still attractive to the Employer, faced with his various alternatives!).

There are other pitfalls for an unsuspecting project manager, who is positioned about midway between Scylla and Charybdis. Some relate to payment for a part of the contract price which is due. This might be sufficient to pay off the creditor who has called for insolvency measures (if there was only one), thus removing the insolvency and removing all authority from the bankruptcy official. Conversely, if he withholds payments due, he may be accused of being in breach of contract, or even of repudiating it, in order to extract the heavy damages his Employer would face. The project manager needs to tread with the greatest care. If he makes a new deal (or advances money) to the official, he may cause trouble with the surety of the performance guarantee. The project manager must be under no illusion about what powers he has, nor the powers of his opponents.

8 *Quality assurance, reliability, the Limitation Acts, latent damage, strict liability and the Consumers Protection Act 1987, occupier's risks and arbitration*

It is a truism to say that engineering projects have recently changed: they always have changed, ever since man first made a spark by striking two stones together or watched a kettle boil; and they always will. However, it is safe to say that the changes which have taken place over the past few decades (possibly as the indirect result of two world wars) have been very considerable by any standards, and it is fitting that we should finish this book with a summary of the chief enactments have brought about by developments in people and things.

The changes in engineering have been legion, and it is quite outside our terms of reference in this book to make any attempt at listing them. They have been caused partly by new and cheaper processes (automation and robotics, for example), new electronics (computers, digital systems, transistors instead of valves, and hosts of other developments in this electronic age), by new materials (such as silicons, plastics, aluminium, titanium, plastic sheeting, oils and chemicals, detergents, etc.) and by the emergence of backward nations, self-government, the spread of industry, growing export trade, and such like. We have also experienced the world getting 'smaller', with international competition and the formation of international groups. These have caused us to look less parochially beyond our own country and defer to the ways and thoughts of others (e.g. the EEC or various trade pacts, etc.).

All in all, we have to take a new look at the classic ways we have so conservatively followed for so long, and perhaps realize today that we must take far greater account of the liking that modern man has for legal assistance against personal affronts. Therefore, in this chapter we shall take a brief look at the different ways in which modern engineering is moving and the laws which have been made to cater for it. So far we have dealt with the more classic case, which all practitioners must master; now we examine more recent developments, which colour the end-product, and the route taken to get there.

8.1 QUALITY ASSURANCE

Quality Assurance is (as its name implies) an activity separate from (but superimposed upon) the normal relationship between an Employer and his project manager with the intention of obtaining an assurance that his employed organizations will produce the quality of work they profess. It has grown in popularity over the past decade or so, as a result of the steadily increasing capital that an Employer has to invest into any new project, the increasing technical sophistication to be found in most modern undertakings and the growing number of new firms and names – hitherto untried – which have sprung up to handle them. One has only to consider space satellite work, automation and robotics, nuclear installations, advanced telecommunications, undersea oil plants and numerous similar processes to understand how Employers and others found themselves to be getting further and further out of their depth. They had to rely blindly on their technical subordinates, and felt a great need for some assurance that the vast sums they were forced to spend for new highly technical processes would produce the results anticipated. Many of the new suppliers (who professed to be experts) were unknown to them and, in any case, had 'an axe to grind' when it came to the Employer accepting their offers. Employers had found they had to accept them at their own evaluation, and pay heavily for the favour of so doing. There was nothing to protect them, even less to show that their chosen contractor had indeed got the prowess he professed. (Maybe he had a good advertising agent, but precious little else!) In the current jargon, then, Employers might be employing 'a cowboy'!

Thus Quality Assurance came into favour. It was foreshadowed by British Standard No. 5750 (Part I – Specification for design manufacture and installation), and was defined by another BS No. 4778, 'All those planned or systematic actions necessary to prove confidence that an item or facility will perform satisfactorily in service'. In short, can a contractor do what he says he can, and will his quality be as good as he claims?

Basically Quality Assurance is a detailed self-examination by the leaders of a firm, that the firm (or a stated part of the firm) has the requisite skills, organization, knowledge and procedures to be certain it can, with complete self-confidence, carry out those functions it claims that it can. Can it so satisfy its clients? It is not required to manufacture anything: it can be a consultant, architect, designer, adviser, computer analyst, programmer, supplier of technical equipment or just about anything else on which an Employer may risk his money. A project manager is himself such a risk, and an Employer may call for some assurance he can do all that is expected of him, before appointing him. Has he the political and technical stature to give his Employer every satisfaction in his complex project?

An Employer can make use of Quality Assurance practices in a variety of ways; he can demand adequate assurance:

- from his project manager (or his project manager's firm) before he appoints him;
- from his project manager's team before he accepts them, each in his own duties, and pays for them;
- from each contractor, designer, consultant, etc. before he employs them;
- from every tenderer answering his enquiries before his project manager is able to recommend him for acceptance;
- from his own organization that it is capable of setting up and operating the new project, by insisting it has an appropriate Quality Assurance plan which can be expanded if necessary to embrace the site works and the new facility.

An organization adopting a Quality Assurance scheme elects to do so at Board level. It appoints a director in charge, and it is under his direction that a hierarchy of managers and inspectors is superimposed on the firm's organization at all levels. Together they devise a Quality System and a Quality Plan which makes certain that each department can play the part allotted to it in an efficient way – and has all the expertise, training, facilities and ability so to do. Here 'efficiency' must include the time factor as well; a timetable and delivery dates given to a client must be kept. It is only then that he can report to his Board that all is well and the potential customer can be given his assurance. For a customer they will produce:

- the Quality Plan they have adopted for their firm: they will add evidence that they have examined their facilities in accordance with it;
- a Quality Assurance Certificate as to their abilities;
- the Quality Control system they operate and all internal reports to help validate the certificate they supply; since this information must go into their internal organizational arrangements in some detail, such information will not be broadcast, but may be restricted to senior members of the Employer's project team affected by the contract.

Under the continuous surveillance of the firm's Quality Control Director, the system will undergo constant updating and review. It will be subjected to what is referred to as the firm's 'audit' to ensure their methods are still effective and bear out the certificate their management gives to intending customers. The director has direct contact with the Board, and his Quality Assurance team carry no responsibility for the day-to-day operation of the departments inspected.

There are definite efforts these days to extend the scope of Quality Assurance and bring the accent more on to the 'quality' aspect. This has been brought about partly by the requirement for greater reliability, and partly by the bringing into force the Consumer Protection Act 1987. We deal with this in greater detail in Section 8.5, below. It will be seen that Part I of the Act complies with the earlier EEC Directive giving manufacturers strict liability for their products, negligence being no longer a factor. The user can now sue the producer direct, even if he purchases the goods through intermediaries.

To make such an arrangement yield the maximum degree of reliability, a given purchase must bear both the name of its manufacturer and its history. In this way, not only can the producer be brought to book if the purchase is found to be faulty, but accompanying items which might be affected by the same source of error can be readily traced and (if bought by the Employer) promptly eradicated from the project by the project manager. The exact means by which all this is to be put into operation is by no means determined, and it may involve bulk tests by recognized national testing installations. The object of the exercise is patterned on the methods presently adopted by the military services provisioning departments, where similar items must be interchangeable, reliable and able to be used immediately after long storage as spare-parts, or during active service use in some far-distant garrison. A strict control is therefore enforced during manufacture, both on the testing of individual components and the careful isolation of them with their appropriate history. The more general application of Quality Assurance that we have already dealt with will not be affected by any new departures.

The functions of the Employer's project manager are not affected by a contractor's network of inspectors and managers. They have no executive powers, but report directly through their own channels to the Board of the contractor. As far as the project manager is concerned, they are purely 'information-gatherers' (or not to put too fine a point on it, snoopers); their object is naturally a helpful one – i.e. to ensure the contractor's party on-site carries out the responsibilities it has undertaken correctly, efficiently and on time. And they have direct access to their management at the top to ensure all their recommendations are put into effect. The project manager is still in sole command and control of his project, with the duty of satisfying his Employer as normal.

In employing a contractor, an Employer may be satisfied with a partial approach to full Quality Assurance from the company concerned, namely the institution of a Quality Assurance system for just those departments and units which are to be brought into operation at the site. The local head-man (i.e. the Quality Assurance site manager) would still have direct access to his Board (to ensure his reports are duly followed up) but he would probably keep in close touch with the project manager at the site, and his findings would reach the Employer through that channel. Delays to the progress of the works might be kept to a minimum if the project manager has first-hand reports on any criticisms the contractor's Quality Assurance manager might have. A contractor's use of plant might be a case in point.

8.2 RELIABILITY

In engineering generally, but mostly in the electrical and mechanical branches, there has been a steady trend over recent years towards the attainment of better reliability. The results have been getting more pronounced and the aims more intense: there is every indication that this feature will become a prime consideration in all future engineering contracts. There is a firm demand by all users for something which approaches the age-long ideal, namely that it not only works when it is installed and passes its tests on completion, but goes on working without a hitch throughout its anticipated lifetime.

The reasons are obvious: what has hitherto been a hope is now a practicable possibility. The needs have become more promising: (a) the cost of maintenance steadily advances; (b) modern developments have brought new materials into regular use; (c) manufacturing processes are more sophisticated; and (d) advances in communications have expanded our normal markets well beyond the boundaries of one's own country to the remote places of the earth. We have met overseas competition and international associations, so that the views of the other man have to be taken into account just as much as our own customers at home.

The days of *caveat emptor* (beloved of our forebears) are gone; the maker is being accorded greater liability for his products. Who can doubt that the constructional and building trades, the most apart and conservative of our engineering industries, will be close behind the demand?

8.2.1 The high cost of maintenance

This is not only the greatly increased cost of having maintenance engineers available: it reflects also the much wider knowledge and advanced training they must have before they can grapple successfully with faults in today's highly sophisticated processes. Increasingly it has become more expensive to locate a faulty component than to replace the misbehaving part by a complete new machine or sub-assembly. Whole factories are being designed on this principle. A family car is a good case in point: here a misbehaving engine is repaired by substituting a brand-new engine or, at any rate, an overhauled and tested one. Maintenance has become a systems problem, and no longer a components one. It is inefficient and uneconomic to adopt the latter.

8.2.2 The advent of modern methods

Major automation schemes, from single machine to complete factories for mass-production, have meant that one simple fault can cause extensive disruption of a whole centre, and lead to much loss of time and money. Faults must not lead to plant breakdown; unreliability is no longer merely a local nuisance, it represents big losses of money and production. The same applies to the introduction of more sophisticated aids to production, such as computers and robots, and there are some interesting side-lights which can be quoted on these topics. Such, for example, as the chemical plant which had been run for years by old experienced men, with greasy hands and years of employment on the plant. They could tell at a glance when things were just a bit off course, and frequently found it necessary to make slight adjustments to the rough tap they had been placed to supervise. Computer control could do the jobs of all of them much better, much cheaper, much quicker and much more diligently, but it had to be turned down on the grounds that there would then be no people available with the experience to run the plant *if the computer system broke down*! Or the maker of inflammable chemicals who also turned down the opportunity of using computer control (with equal promises of the fruits of modernization) because there would no longer be men on the site to man the fire-brigade appliances in emergency!

It is no coincidence that both of the above examples stem from considerations of computer use, because the main force behind the whole demand for greater reliability springs from computer activities. Although their operation had benefited greatly from the demise of the thermionic valve in favour of semiconductors, and the substitution of miniaturization for the use of individual components, it was still inherent in the functioning of most of them that they never lost their programmed memories – i.e. they were *never* switched off or broke down. It was this need for constant operation which started a demand for increased reliability. The use of programmed devices is now so widespread that using an equipment which runs 'for ever' has become both a habit and a necessity.

8.2.3 New materials

Probably accelerated by two world wars (when development work was undertaken almost regardless of the cost, in terms of money), the past few decades have had available many new materials, which have made efforts at increasing the reliability more than just a passing hope. Silicones have given greater durability, greater moisture resistance and higher temperatures of safe operation; plastics have given greater castability, added strength, resistance to shock, more uniform sheets and new materials for bearings (nylon); modern painting materials are more durable, wider in their ranges of application and less poisonous; lubricating oils are vastly different from their equivalents a few decades ago; and so we go on. Without all these, any thoughts of adding to our reliability would have been largely theoretical.

Perhaps the best home-spun example of the way things have improved over recent years is afforded by the family car. It is not so long ago that all engines had to be decarburized, and cylinder valves ground in snugly every thousand or so miles. Radiators were fitted with warm 'muffs' in winter, and needed topping up weekly; fresh anti-freeze each winter was drained off as soon as possible when winter was barely over. A couple of dozen bearings were fitted with grease-cups or oiling-points and had to be treated every thousand or so miles. Ignition circuits had to be tuned and adjusted frequently, and even the dashboard clock had to be wound up and set once a week. All this has changed. Engineering projects have changed accordingly, and reliability has to be measured in the freedom from regular maintenance and absence of faults which have been achieved.

8.2.4 Exports

The final consideration is the export trade. It is one thing to have to carry stocks of spares readily available to the home market, but something else to keep a machine running without trouble in the Kalahari Desert or the jungles of central Africa, remote from spares or machine shops, and with no expert attention apart from what one arranges for oneself. Above all, users want complete built-in reliability under all conditions and weathers.

The world has become effectively smaller. We have joined our friends in 'associations', such as the EEC, and we meet our competitors wherever business is to be found. We must pay attention to what they tell us, and what they can offer against us. The story they all tell is for increased reliability in the field. Product liability of the machine manufacturer has become the norm, tests on completion or at the factory are only part of the history of the product. Contractors' promises, assurances and guarantees, are still met with disbelief. Hours of constant work on the job, backed by such statistics as MTBF figures (Mean Time Between Failures) are what a successful export policy demands.

A project manager would do well to bear in mind these modern requirements when formulating his designs for the project, and also to ensure that his contractors bear them in mind when filling in their details or carrying their designs into practice.

8.3 THE LIMITATION ACTS

These so-called 'Statutes of Limitation' sought to set a limit on the period during which a contractor might be at risk from a defect he had introduced into some contract works at an earlier date. Inside the period he could be sued for breach of contract, a matter against which he was wise to effect insurance indemnity. Today it has all become much more complicated.

The Limitation Act 1939 aimed to clear the matter up, and set the extinguishing period at 6 years for a simple contract (under signature), and 12 years for a contract under seal. After these periods, the contractor was no longer at risk, as there was no legal remedy either under tort or for breach of contract. Shortly afterwards, it was realized that with some sorts of injury to persons it would be better to have a shorter limitation period. There must be sufficient evidence, both that the physical effect was serious (how serious?) and that it had been caused by a defect in the works. Some illnesses took a long time to become serious.

The Limitation Act 1980 therefore altered the limits when bodily harm was involved to 3 years, but this could be extended further if it could be shown that there was insufficient knowledge available to support a claim: (a) that the plaintiff had suffered danger to his safety, (b) suffered actual damage to health, (c) its seriousness (so that damages could be assessed), (d) that his illness was likely to be permanent and (e) that it was indeed caused by the alleged defect in the contractor's work.

This was all very well as far as it went, but it did not consider the circumstances when a defect was out of sight and did not become manifest until some time after the defect had been caused. Also a builder could argue, for instance, that he was merely doing what had been ordered and the fault lay previously with the architects or the engineer's design. Who owed a duty to whom, and what constituted negligence? When did the periods start to run?

There were several such cases around 1970, some of which eventually went to the House of Lords for a decision; and these established the following provisions for the construction and building industries:

- Councils and similar bodies owed a duty to the public when they passed plans and specifications which did not comply with the building regulations.
- Their servants were negligent, especially if their inspections of foundations were inadequate or if they passed work as 'good' when it was, in truth, 'bad'.
- A cause of action accrues not at the date of the negligent act or omission, but at a date when the plaintiff first observed the results of the defect (or should have observed them if he had been reasonably alert).

That is to say, time does not begin to run out until the plaintiff discovers the damage has been done (or ought with reasonable diligence to have discovered it): 'It is a greater hardship on a purchaser of property when damage only becomes clear after an artificial expiry date, than on a builder, designer or engineer who finds himself sued many years after he has left the site. The householder would be without remedy, and could not have discovered the fault in covered-up work any earlier. The period of limitation may have to be postponed indefinitely.' This was tested in a case that went to the House of Lords, when there was a unanimous decision that:

- Local authorities exercising their functions under the Public Health Act may be liable if they inspect the foundations negligently or fail to see that the regulations are complied with. They must ensure that non-compliant foundations are not covered in, and must make physical inspections in so far as this requirement necessitates them.

- An owner or occupier will not be statute-barred if defects only become apparent more than 6 (or 12) years after the negligent treatment by the council.
- Until the condition of the property gives rise to danger to the health or safety of persons on the premises no breach of duty has taken place.

8.4 THE LATENT DAMAGE ACT 1986

Theoretically no building, however old, is excluded from this ruling and the construction industry must remain liable in negligence long after the builders have quit the site. It was this open-ended liability which faced the draftsmen of the Latent Damage Act 1986, to settle in a legal way the dispute between the builders and the occupiers of the property. Their solution takes effect primarily by amendment of the 1980 Act: the latter dealt with latent personal injury limiting the period of making a claim to 3 years, and thus no new provisions were needed for personal injury in the 1986 Act. The 1980 Act was taken largely as the model for the 1986 Act (but it excludes damage caused by personal injury). It is hardly surprising that, with two conceptions so widely different, the legal officers of the Crown had the most unenviable task of trying to satisfy both and, at the same time, to give their work the legal clothing such an Act requires. They managed to satisfy neither party completely, and many matters are still open to interpretation by the trade, by the courts and by the insurance companies. As, it stands under the Latent Damage Act 1986, the builders' liability may drag on for 17–18 years after they leave the site. The occupier may change several times from the one who was the original party to the contract which produced the faulty work. They may be only third parties, having no knowledge of the early history of the premises concerned. Nevertheless, they can still sue regardless of the time at which they themselves took possession, provided that they are not excluded by the efflux of time since the fault was produced. These periods of time are illustrated by the chart in Fig. 8.1: the Latent Damage Act 1986 is now the law on the subject, however displeased the two sides may be about it.

Similar actions for negligence (and, of course, actions for breach of contract, solely reserved to the two actual parties) may be brought against designers, solicitors, consultants, accountants, etc., as well as

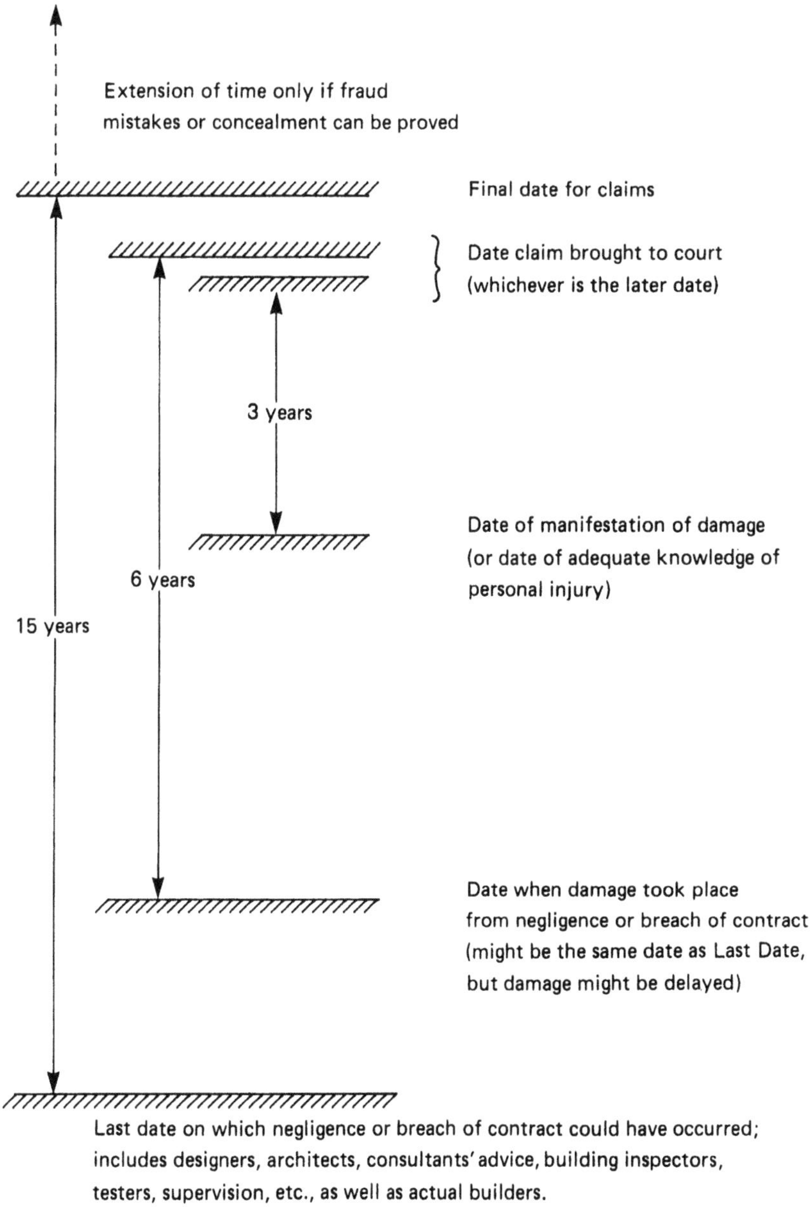

Figure 8.1 The Latent Damage Act 1986

architects, contractors or local councils, but the last-named are only concerned if damage has occurred (or is likely to occur) to the detriment of public health and safety. Briefly, the 1986 Act allows a special 3-year period after a case of latent damage *appears* subsequent to the date of the occurrence itself, but with a 'stop' at 15 years after the last date of any action or negligence which led to the defect (unless the plaintiff can show there has been fraud, mistake or concealment). The start of this period of 15 years is a matter for examination in each and every case. The negligence may have occurred due to a mistaken interpretation of a contract specification – i.e. it should have been noticed and corrected at, or prior to, the Tests on Completion, making them the starting date. Alternatively, the contractor may have had an extended duty during his warranty period, but usually he is only obliged to put right things brought to his attention during that period – in which case, the last opportunity might have been at the end of the warranty period. Or the fault might be proved to lie in an earlier design, or an error by a consultant or architect, which the builder had no reason to question, and the design was put into operation with nobody realizing the rules were being broken. Trouble must eventually occur. Finally, of course, is the usual case, where nobody is aware of the defect until its actual results break surface many years later and cause danger to the present occupier.

The Act depends on three dates:

- The last date when any act or omission amounting to negligence occurred, that the damage complained of was attributable to it – i.e. the last date of a breach of duty.
- The date on which a cause of action accrued (an affected person cannot start an action about some thing he doesn't even know exists).
- The date on which there is sufficient knowledge and appreciation to allow a sound action for damages to be started.

The first two dates are not new to the law. They must be fixed in consideration of each case that occurs, and the periods of limitation which apply to the damages envisaged. The starting date (or date of adequate knowledge) as applied to ordinary damages (not affecting a person's health) is a new conception, introduced and defined by the Latent Damage Act 1986 itself.

To sum up, the time limits for an action to start in respect of damage NOT INVOLVING PERSONAL INJURIES are whichever expires the later of:

(a) 6 years from the date on which a cause of action accrued; or
(b) 3 years from the starting date (i.e. the date of 'sufficient knowledge');

this date can, however, be extended if the person concerned is already disabled physically or mentally; but

(c) in any case, not later than 15 years from the date that negligence occurred, unless there can be shown to have been fraud or mistakes or concealment. In such a case, the 15-year period does not begin until the fraud, etc., was discovered (or could have been discovered with reasonable diligence).

Note that as the claim is for 'damages' at tort, actual damage of some sort must have occurred before any 'damages' can be assessed. What is claimable is the assessed value of such damage as has actually taken place. There is no such claim for something that 'might cause damage' (i.e. for an actual defect *per se*. Latent damage involving any form of personal injury is already dealt with by the Limitation Act 1980, which is not affected by the new law: the latter brings non-injurious damage into line). In effect, the plaintiff has a period of 3 years to prepare and bring his case before the courts, after he has found out what is happening. He has longer only if his legal advisers say it is necessary.

What has been said so far applies only to the UK (there are special differences to meet differing Scottish law). Project managers must pay especial attention to the periods of limitation (if any) which are imposed on their contractors overseas when the site and the location of the works may be under a different jurisdiction. The rules will almost certainly be different from those ruling at home, especially in view of the changes wrought by the Latent Damage Act 1986, and it is probably under local laws that any claims will be made, regardless of which law is selected to apply to the contract itself. The terms of his contracts must allow for extra insurance if the risks are more prolonged (the more so if they are open-ended!). Remember, damage might be caused by any of the persons concerned with the basic designs, and not by the builder who happens to be on the spot: he might be able to plead not-guilty to local charges of negligent work if he can demonstrate he was merely doing what he was instructed to do, and the fault lay with the design team who first conceived the idea – and the project manager!

8.5 STRICT LIABILITY: THE CONSUMERS PROTECTION ACT 1987

The idea of strict liability has been around for a long time, and some nations (such as the USA and some European countries) have adopted it.

Hitherto in the UK there have been only two ways in which a 'damaged' person could obtain redress:

1. If the person had a contract with the guilty party under which the latter failed to carry out an action he had undertaken to do, he could be sued for 'breach of contract'; by the 'Rule of Privity' of contracts, a person who had no such contract was not allowed to use this process.
2. If there was no contract, any person could sue a guilty contractor, for example, at common law, under any law of tort, *provided that he could prove*:
   (a) the guilty person owed a duty of care to the person concerned, or to the public generally (which included the 'damaged' party, of course);
   (b) that he did not carry out this duty or was negligent in doing so (either by omission or by lack of effective action);
   (c) damage was suffered as a direct result of such negligence; the damage suffered was a foreseeable consequence of the negligence complained of, by any reasonable person;
   (d) the estimated cost to himself of the damage or injury he received resulting from the defect.

Thus a contractor working on a nearby roof has a duty of care to the public using the pavement below. If by error he omitted to fit a kick-board and an employee dislodged a spanner, which fell on to a passer-by and injured him, the passer-by could sue him for negligence in tort. The contractor could have foreseen such an accident happening, and he was negligent in not taking adequate steps to prevent it. An employee could sue his boss (assuming his contract of employment did not cover the point) at tort if he issued him with defective equipment to work with. The Employers Liability (Defective Equipment) Act 1969 dealt with this type of negligence, and made it unnecessary to prove a definite duty owed to the employee.

Much of the earlier legislation was brought together in the Health and Safety at Work, etc. Act 1974, which we deal with separately in this chapter. Although basically a part of the criminal law, it can lead to a common law liability in such a case as we have described, above. Many people have thought that it should not be required of an employer that he is constantly 'looking over his shoulder' to make sure all the things he uses in his work are exactly what they are purported to be, having no unnoticed faults or blemishes. He (the contractor) was liable at law, and he could not pass the liability on without thorough examination and tests. There was a need for a 'strict liability' that when a supplier or maker sold something, it was just what he said it was and had no faults or blemishes in its construction. The maker should shoulder the responsibility, irrespective of whether there was an element of negligence or not. Two bodies in the UK have studied the matter: (a) the Law Commission, which made public its report in 1977; and (b) a Royal Commission, which made public its report (the so-called Pearson Report) in March 1978. Both bodies recommended that a principle of 'strict liability' should be adopted. Simultaneously, the European Economic Community was considering the situation as it had been faced in some of its member countries. It aimed to collect together the important points of the member states, and in 1979 produced a Draft Directive, which it discussed with them. Some found it far more stringent than anything they had used hitherto, and it was six years before a common plan was produced. Eventually, on 25 July 1985, the EEC sent out its Directive on strict liability for ratification and adoption before 31 July 1988 (Ref. 85/374/EEC).

In the UK this has resulted in the drafting and adoption of the Consumer Protection Act 1987, of which Part I (together with Schedules 1 and 3) has met the requirements of the EEC Directive. The remainder of the Act deals with the position as it affects consumer goods, such as are used in the private home, and has not been separately analysed for the present book which deals exclusively with the engineering industry's contracts. The new statute deals only with the strict liability of the producer; matters of damages and negligence have to be dealt with by existing procedures, using existing Acts and their associated case law. These limit neither the total amount that can be approved as damages nor the number of co-existing claims which can be made to the courts in respect of the same fault.

In what follows we examine in closer detail the contents of the new statute, which introduces a new liability to those already existing. Its main difference is that negligence is no longer required to be shown by the aggrieved person: he merely has to show that there is a defect in the product, and that this defect caused the occurrence which led him to claim damages. The producer is then liable automatically (subject to certain matters he can call in his defence, listed in section 4 of Part I of the Act as being acceptable arguments).

The Act is divided into five parts, and Part I is stated to make 'such provision as is necessary to comply with the Directive of the Council of European Communities, dated 25th July 1985, on the approximation of the laws, regulations and administrative provisions of the member-states concerning liability for defective products'.

It applies to all products, irrespective of whether or not they are intended for private or commercial uses. Its terms include water, gas and electricity (a note is added that it is not envisaged that 'defective electricity' might be produced, but faults might occur in the generating and distribution machinery used!).

A 'defect' in a product or material is defined (for the purposes of Part I of the Act) as causing measures of safety not to be such 'as persons generally are entitled to expect'. Here the word 'safety' includes risks not only to persons themselves (i.e. injuries or death), but also to the things they own (i.e. land, articles and property of one sort or another). Claims must take into account:

- Any limitations in use prescribed, or warnings against specific misuse given with the product.
- How the product was actually used, and period for which it had been in use.
- Any likelihood of defect arising during course of transition between the maker and user (e.g. any packaging matters, treatment or adjustment or testing applied by intermediaries).
- The time factor, that is the time of production and the time of damage must not be unreasonably remote.

The persons liable are:

- The producer or such person as by putting his name or trade mark on the products holds himself out as the producer.
- If made overseas, the importer in the course of his business (i.e. it does not apply to an individual who happens to bring the product into the UK himself as a personal possession).

The term 'producer' of a product can include the manufacturer of a component part or of raw material, or of a finished product, or a person who applies an industrial process to a 'finished' product; the choice (if one has to be made) rests with the aggrieved person. Retailers (or the second party of a sale/purchase contract) are not involved in Part I of the Act except in so far as they refuse to respond to a request from the aggrieved party to identify the source of goods claimed to be faulty and leading to a claim for damages. If two or more parties are liable for the same item of damage under Part I of the Act, their liabilities are both jointly and severally.

Liability under Part I for damages follows automatically if a defect in a product is proven – i.e. it is 'strict liability'. The recovery of damages otherwise follows the application of existing laws subject to section 5 of the Act. A defective part used by a contractor may form part of the works of a contract, and then can often be replaced under the contract terms; if no contract exists, an action at tort (civil law) is the only procedure using the new Act to allot liability. The other parts of the new Act (Parts II–IV) apply only to 'consumer goods' – i.e. 'goods ordinarily intended for private use or consumption' (section 10(7) of the Act). Recovery of damages may then follow a separate procedure (sections 10(6) and 20(4).

The Act is to be read industrially in conjunction with the Health and Safety at Work, etc. Act 1974, which lays down the liability for employers and contractors to take all precautions necessary to preserve the safety and health of all people concerned with, or in the area of, the site. The new Act adds a further definition of liability when it deals with faulty products or machinery. The defendant under the Health and Safety Act is entitled to plead that he accepted what appeared to be a satisfactory component, but which after examination and use proved to be faulty. He was not negligent in using it after making a reasonable examination, as under the new Act, the producer was liable for hidden defects.

As regards the periods of liability, the Act is, of course, subject to the other current Acts such as the Limitation Acts of 1939 and 1980 and the Latent Damages Act 1986.

8.6 THE HEALTH AND SAFETY AT WORK, ETC. ACT 1974

Although this is not a new Act, it is included here as it represents the ethos of the modern working site, namely that everybody (not only the contractors themselves, but their employees and the site owners) have a duty to their 'neighbours' on the site to do all they can to make sure the working conditions are as conducive as possible to safety and good health.

Unlike most laws, this Act is not directed primarily at the malefactor, to ensure he pays for any damages his waywardness produces, but rather it is a criminal law which makes it a criminal act to put people at the workplace unnecessarily in danger. On indictment, a person can be treated as a criminal, and given punishments up to imprisonment. Unlike the normal criminal law, it is not assumed that a person is innocent until the Crown proves him guilty: he has to prove (when injury or damage has occurred) that he took 'reasonable precautions' against it happening. He must show that the precautions he took were both 'reasonable' in the practical circumstances existing at the site; and that he both took them and was prevented by impracticability from doing anything further. If he is unsuccessful, he might in certain circumstances, be awarded imprisonment for up to 2 years.

The Contractor has, of course, the greatest responsibility to make sure of the safety and health of the employees he engages for his contract work. Under section 2 of the Act, he must have a general policy on the whole subject and must provide a written statement of the organization and requirements he needs to carry out his duty of care for his workpeople. This he keeps up to date (and rewrites it as necessary to fit the particular site he is working on), but he must also bring it to the notice of all his employees. They themselves have a duty to take reasonable care to avoid injury and to co-operate with their employers: they must not vandalize, remove or tamper with anything provided for their protection (sections 7 and 8 of the Act). The Act also imposes a responsibility on all others who may be concerned (section 6), so that designers, importers or suppliers of articles or materials to be used on the work do not make them in a form which would add to the danger or unhealthiness of the workpeople using them. There is, of course, a similar requirement from the employers that the tools, machines and methods they cause their workmen to use must not subscribe to their lack of safety or health.

Naturally, the project engineer is similarly burdened. Not only does he employ direct labour, but he has probably directed or approved the methods which his contractors propose to use in carrying out their work on-site. Some of the precautions, such as those against fire, will be wholly site arrangements which he has made, and which his contractors assume are adequate for the dangers likely to exist. Each contractor is responsible for his own works and workforce and only the site manager (and possibly the project manager, or possibly his constructional engineer on the site) has the powers to see that such arrangements are both adequate and operationally sound. Finally, of course, he alone is responsible for the application of the Act to matters which might fall between the different contractors working simultaneously on the site, to the areas where their two or more responsibilities overlap and it becomes the business of none of them. There must still be a coordinator. The Employer's responsibility may fall on the shoulders of his appointed management, and it is up to them to see that his responsibilities are carried out; it may be necessary, for example, for a general policy statement (such as we have referred to above) to be produced and circulated in respect of any joint (or site) arrangements which the project manager makes for the safety of all contractors operating on the site. The mere fact that he can bear such a personal responsibility is a special inducement to him to take all care and precautions, and not leave them to somebody else.

Although the Health and Safety at Work, etc. Act 1974 is basically a statement of liability under criminal law and thereby a criminal statute, it can nevertheless lead to a civil liability in which fault and negligence play an essential part. Negligence is an essential element of a claim under tort generally – i.e. a claim at common law. The recent Consumer Protection Act 1987, which we deal with elsewhere in this volume, introduces (at Part I) the principle of 'strict liability' for all those who put products into circulation (i.e. their designers, manufacturers, importers, installers, and the like). This new Act introduces the principle into English law – mostly at the behest of the European Economic Community, of whose Parliament the UK is a member state. Strict Liability is already a legal conception in the USA and, to a greater or lesser extent, in a number of European countries.

There is obviously a close relationship between the rules introduced by the Health and Safety at Work, etc. Act 1974 and the more recent Consumer Protection Act 1987, such that Schedule 3 of the latter is entirely devoted to linking the two statutes together. The differences are, to a large extent, caused by the addition of a 'fairground' to the conception of a 'place of work' and the duties of employers under the Health and Safety at Work Act are not much affected. The main difference is that while under the older Act people must take 'reasonable precautions' to avoid the stigma of 'negligence', under the more recent imposition of 'strict liability' they are sometimes to blame, even if no especial negligence is proved.

Finally, we might add a note about the periods of liability during which a person is 'at risk'. These are determined by the Limitations Act 1986, or if physical injury is caused, the Limitation Act 1980. For the recent 'strict liability' Act the limits have become:

For bringing any action against a producer or importer 10 years from the time the defective product was first supplied to a customer.

For any action resulting in a claim for damages: 3 years from the date when a cause of action first occurs. This period can be extended when the available knowledge at the last date is inadequate to further the claim.

The latest word on this is probably the Latent Damage Act 1986 to which reference should be made.

8.7 OCCUPIER'S LIABILITY TO VISITORS TO SITE

Every occupier of a premises has a duty to safeguard third-parties visiting the premises, and the Employer is no exception. He exercises his duty through his project manager, who is therefore personally responsible to his client for the safety of visitors on-site. Note that the duty is expected from the occupier, not the owner, the lessee, the absent landlord, or such like, but the person who is *occupying* the site. If any accident occurs by which a third-party is injured, the occupier can be jointly charged along with the people who actually caused the injury or damage. The charge is negligence, in that the necessary precautions weren't taken to ensure the third party's safety. A contractor, his employees and his visitors, observers, callers, etc., are all treated as third parties, and none of them is likely to be the occupier.

Therefore, in arranging a contract, the project manager should see that it contains a clause which indemnifies the Employer from all costs, damages and charges which he may sustain from claims made against him by any visitor or employee on the site to the extent that the Employer was not guilty of actually subscribing to the damage the third party sustained. The exclusion cannot, of course, remove from the Employer's staff on the site any direct blame they might incur through their own actions, but it can relieve the Employer from a claim made against him, as occupier of the site, when he has been blameless for the actual accident producing the casualty.

How is an occupier defined? He is the person who controls the premises, and especially the access of visitors and other entrants. He is the person who can say: 'Shut that gate and keep them out', or 'Open it up and let them in'. He need not be the owner, who might be miles away, having handed control over to the project manager. The occupier owes his same common duty of care to all persons he allows to enter, and hence can include his contractors, their staff and employees among his 'visitors'.

He can remove his duty of care either by refusing callers their entry to the site, or by notice delivered to each of them in person, either verbally or by a notice saying so and placed in a position in which it can be seen by all visitors entering. For example, a clear warning placed near the entrance gate telling arrivals that the site is pitted with deep holes which they might fall into if they leave the clearly defined and marked path, would exempt the Employer from damage sustained by a person leaving the path and falling into one. A similar duty of reasonable care must be afforded to the police, firemen, ambulance men, and so on, who have cause to visit the site in the course of their duties. It is likely all project managers have seen the warning notices to firemen concerning coloured signs outside buildings, and saying they must be switched off . . . here . . . before playing water on the building concerned. The principle is the same.

The duty of care must also be extended to children and trespassers if it is known that they are prone to come to the site and have ready access to it. A properly fenced site would be adequate precaution, if it really did keep people out unless they used extraordinary methods not to be anticipated. People on their way home from a local factory were wont to cross a site and some were eventually knocked down by a shunting engine. Although they were trespassing, the law held that the occupier was to blame, as their practice was known to him beforehand and he did not take reasonable care for their safety.

What is 'reasonable' care? This is ultimately a matter for the courts to decide, but generally it consists of an effective warning and adequate measures to prevent any user of the site (trespasser or not) accidentally getting into a situation of danger. The key-word is 'accidentally'. He must not absent-mindedly (or if children, at play) get into such a dangerous situation, even if he disobeys a 'keep off the grass' notice. However, it is not incumbent on the occupier to guard against people who intentionally avoid safety precautions, or employ special tools, e.g. wire-cutters, scaling ladders, parachutes, etc. That would be considered 'unreasonable' by any right-minded project manager. You can never keep a determined entrant out. The courts take into account all the relevant factors, including the size and wealth of the people concerned. They will not expect a small firm of limited resources to undertake such exotic precautions as (say) a national body such as a railway (where the danger can 'creep up on one' almost silently). They will be expected to take less 'risks' than a small local owner, and employ greater resources so to do. Thus clear warning notices at strategic points or at frequent intervals, well-maintained fences (even if only placed around actual danger-spots), patrolling watchmen and visits to local schools all help the occupier to create an atmosphere of 'caring'. The employment of competent contractors (no 'cowboys'), adequately lit pathways and reasonable anticipation of illegal entrants all play their part.

The titular occupier is the Employer. It is his site, and he is erecting his project on it; he will expect his project manager, as a part of his duties, to protect him from costly claims. In certain circumstances, the project manager himself might be held personally responsible, in place of a distant Employer who never visited the site. And do not overlook the intrinsic danger of partly built contract works, left overnight by a thoughtless operative or foreman; they can be just as dangerous an attraction to unauthorized children as anything else.

8.8 DISPUTES AND ARBITRATION

All valid contracts are supported by the law, that is they are backed by court rulings. It must, however, never be overlooked that such hearings cost both parties thousands of pounds (nay, hundreds of thousands of pounds) each day by the time one has paid for learned counsel, their juniors, 'sweeteners' and solicitors (and with, each one, his time, spent on briefing, preparation, writing opinions, and the like). Many disputes are just not worth taking a risk over at law, especially if there is any possibility that the finding of a court might go against an Employer. It is not surprising therefore that most contracts (and the Forms of Contract or Standard Conditions of Contract on which they are based) contain agreed terms that a dispute shall be given prior submission to some cheaper and quicker form of settlement. They are trying thereby not to evade the law, but to provide an unbiassed decision without the attendant high costs and delays which a court action would involve. They often envisage the following.

(a) An initial reference to the most unbiassed man on the site, namely the project manager (or the Engineer, if so named in the contract itself). Although he has been appointed by one party (the Employer) to control his project, he is usually recognized to have a second role as quasi-arbitrator when any dispute arises between the parties to the contract. If he is an independent member of a firm of consultants, his authority is thereby enhanced. He is expected to be guided by the ethics of his profession, and to give an unbiassed third party view of the matter in dispute. The two sides assemble their documentation and make their cases: he makes his decision on these and on his general knowledge of the local situation, gained from his daily contact with the project and the parties' intentions. His rulings are immediate, local, cheap and, in general, acceptable to both parties.

(b) If (a) fails, a second stage is by the process known as arbitration, the determination by an outside person or an assembly of persons (arbitrators), not necessarily lawyers or connected directly with the law. If the parties concerned are all in the UK, the hearings are held under the Arbitration Act of 1950 and 1979 (or any statutory modification thereof). There may be one arbitrator or three, and the proceedings are held in private. We shall deal in greater detail with arbitration below. With parties, some of whom are overseas and not subject to UK laws, arbitration is usually arranged under the Rules of Conciliation and Arbitration of the International Chamber of Commerce, and the seat of such arbitration and the appropriate laws must be agreed between the claimants.

(c) The only other remedy is at law. Indeed if the subject of the dispute is an overwhelming one and involves considerable reference to the law itself or to detailed accounting procedures, it may be much quicker and even cheaper to go direct to law and cut out the arbitration procedures. Not only are the assemblies of counsel, solicitors, and the like, just as formidable, but the arbitrators' time does not have to be paid for separately! The Arbitration Act 1979 completely revises the relationship of arbitration queries and the law. The onus is on the parties to the dispute to seek legal interpretations, either before or during, or after, the findings of the arbitrator. Meanwhile an arbitrator will proceed with his hearing of the case, and make his own legal interpretations.

An arbitrator may reach his decision on documentation put forward by the two parties when they present their cases at a preliminary hearing. On the other hand, he may not be satisfied with the documentation alone and may require a full series of meetings before he can fairly assess their relative merits. At these subsequent hearings, personal evidence may be called for, each witness being subject to skilled cross-examination by professional barristers. They will probe and nag at anything non-factual, indecisive or glib in order to discredit the witness or to break down any points in the evidence their experience leads them to consider dangerous. Technical points may call for the attendance of expert witnesses. The hearings may last a considerable time to examine all the points the arbitrator is doubtful about, and the results depend a great deal on the time and expense devoted to preparation of the case. Therefore, with complicated matters, it is better in most cases to accept similar expenses and go direct to the courts and proper legal hearings.

It is in the more minor disputes that the arbitration proceedings come into their own. Often a 'settlement out of court' may be offered at the arbitrator's preliminary hearing, once the parties become fully aware of the strengths and weaknesses of the rival claims. Both parties must consider their own cases honestly, and decide before coming to the meeting, at what level they will accept a settlement if offered. The arbitrator must, of course, agree to any proposals made, but if these are put forward seriously and are patently reasonable, he is not likely to obstruct them. Full allowances must be made for the uncertainties of the future and the additional costs which are sure to be met and not recovered. A settlement at this stage is not to be taken as a sign of weakness by the opposition: it is just good economic sense. Usually it will mean both parties have to give way to some extent from their prepared positions. Rarely can they expect to get '100% with costs', however strong their case, and what they are offered must be compared with what they might hope for if they went ahead with their claim.

Although arbitration is not a process of the law, it is subject to certain laws, and the courts will uphold its findings (if they are given in writing – as they almost always are). The award of the arbitrator is final, though certain facts can be raised with him direct if it is considered he has given insufficient value to them. The courts will enforce his findings (if applied to), and these can also be used as a basis for further actions (e.g. in claims for damages, breach of contract, etc.). An arbitrator is usually required to make his own interpretations of the law as it stands, but the parties to the dispute can agree to state a legal query calling for a formal decision by the courts on any interpretation of the law or of the contract itself, and (if necessary) the arbitration is reviewed.

Panels of suitable persons, to act as arbitrators, consist of both legal members and those skilled or experienced in the various disciplines of engineering. The arbitration terms which parties agree when signing a contract usually specify that they will agree a person between themselves, or if they fail so to do within a specified period, they will accept a nomination made to them by the president for the time being of an appropriate engineering institution. He will, of course, nominate someone of his choosing who is best qualified in the technicalities of the matter in dispute.

The person chosen may also have practical experience in what is usually a common rule in similar circumstances in the trade. A choice of arbitration as opposed to going direct to the law is usual when a dispute is of a factual or a technical nature, for example, whether a certain feature which the project manager insists on having is implicit in the contract as signed, or is to be included as an 'extra' for which a contract variation is to be given and an extra price charged. There can be a number of advantages to both parties in such disputes:

- Proceedings may be much less expensive (but they may still be costly!).
- Proceedings may be speedier, indeed on small matters which can be determined from the documentation and without elaborate meetings, very speedy. Delay to the project may be minimal.
- An arbitrator with special expertise may be chosen. This can be of inestimable value in saving time; presentation of a case can be much more readily made and with greater conviction than in a court of law.

- The time and place of the preliminary hearing (and, for that matter, any subsequent meetings) can be chosen to suit the parties in dispute equally with the arbitrator himself.
- All anti-bureaucratic features can be dispensed with and the proceedings given a more 'concerned' approach, as opposed to the impersonal appearance of a public hearing in a court of law. This might be of considerable importance when the parties are still having to meet under everyday work at site.
- Undue publicity can be avoided: hearings are held in private.

It does not take a judge to decide whether an overhead walkway should implicitly be fitted with kick-boards, or a works canteen should have tiled walls and not 'painted' ones. Should handrails be galvanized or just painted? It is left to a third-party expert in such matters to decide just what the contract says or implies.

As to work at site, this must usually be carried out to the full satisfaction of the project manager by the terms of the contract. Acceptance on completion will not be given otherwise. To enable this to be done, the contractor will normally be required to accept *pro tem* whatever ruling the project manager gives, and to take up his rival claim as a separate issue when the project is completed. Much will, of course, depend on the nature of the dispute requiring settlement, but the aim must be to get the project completed on time. After such a lapse of time the subject of the dispute is seen in its right proportions, and the question is frequently avoided.

8.8.1 Discovery of documents

Thus it is to the common good to achieve a means of getting a proper decision from one individual, without the need for the complications which inevitably follow when any question of the law becomes involved. Each party must submit its full range of documentation (even internal memoirs), which reflect on the dispute, at the arbitrator's preliminary hearing, and these documents are then made available to both parties by him. Each side knows the strength of the case of their opponents and can appreciate its weaknesses. Common ground is established prior to the hearing. The chances of a fair award are increased.

This procedure is known as the 'Discovery of Documents', and it is a common feature of all arbitrational or legal proceedings. The objective is to have the parties put up evidence solely on salient features of the dispute and not waste time dealing with matters which are agreed all round. Parties are prevented from springing surprises on each other which may 'fog the issue'.

Oral hearings should always be avoided, if possible: they are costly and lengthy. The legal profession prefer them, especially in the case of non-technical disputes, as they can then manipulate their case after they see which way their opponents are moving, and what points of evidence they are obliged to counter. If one party to an agreed arbitration tries to 'cut the corners' and go directly to a court of law, the other may apply for the case to be held back. The court may decide, however, to continue the action brought before it, especially if questions of legal interpretation are primarily involved. These would be outside the powers of an arbitrator anyway. Or it could agree to hold the case back pending the resulting findings of the arbitrator.

The details we have given so far apply to a situation in the UK and English law; they may not apply in countries overseas, and a project manager, faced with a foreign dispute, must consult the local legal authorities as to his position in an arbitration before he makes any final decision as to the line he should adopt. Some countries, for example, have a much less satisfactory system of appointing their arbitrators. They recognize only a court of three individuals. Each party to the dispute separately appoints one arbitrator, whom it selects without reference to the other; the two thus chosen then appoint the third. A party thus has unlimited opportunity to hold up the proceedings of the court, or even the appointment of the third member, through the procrastination of its appointed member.

A project manager should find out what then happens and how a court settles a case if the decision is two to one. In some countries the findings of the arbitration are themselves binding, and are not subject to any scrutiny or correction which the law of that country could provide. There are other differences in detail, but these can only be described as matters of local information as and when the occasion arises. It sometimes happens when a court of arbitration consists of three arbitrators, that the findings of any two constitute the award of the court, which is binding on both parties. If, however, the third member is appointed as an umpire, he will decide individually, any question put to him by the two arbitrators on the court, and his individual finding is the award whenever such question is one which his fellows have disagreed on.

8.8.2 Costs of arbitration

In whatever way the disputants submit their dispute to an award of arbitration, considerable costs can be involved, and the arbitrator or arbitrators have the power in the UK to decide which party is to pay the costs. These are often allotted against the losing disputant, but the winner might also be considered partly to blame and is then ordered to bear a portion of the costs.

Costs are considered under three headings, and different rulings can be made regarding each of the three; they are:

- *Cause*. These are the costs, which each of the parties has had to assume up to the time of submission of the question to the arbitrator. They were probably assumed without reference to the other party.
- *Reference*. These are the costs, each party working separately in connection with the preparation of their cases for presentation to the arbitrators, and include solicitors, advocates, printing, etc.
- *Award*. These are the total charges of the court of arbitration itself, including the fees of the arbitrators, any technical advisers they consult, the hearing-room hire, secretarial, etc. Basically they were incurred in equal parts on behalf of both disputants.

8.8.3 Award

The arbitrator(s) will hand down their award in writing simultaneously to each disputant. The document must be signed by them all and dated. It must be careful to include specific answers to each question posed (which, it will be remembered, must also be in writing). It must also deal with the question of costs, stating precisely what each party must bear.

8.9 THE INTERNATIONAL CHAMBER OF COMMERCE

It may frequently happen (and especially so if the two parties to a contract are in different countries) that agreement will not be possible for arbitration to take place under the auspices of English arrangements and in accordance with the various statutes of English law which govern arbitration, as it is practised in this country. The law applicable to the contract may well be that of some other country. It is to meet circumstances such as these that the International Chamber of Commerce will settle commercial disputes through its own arbitration courts, organized from its headquarters in Paris. The system is widely recognized and accepted throughout the world.

A typical clause invoking such a court of arbitration would be as follows:

Any dispute arising between the parties to this contract and concerned with this contract, which cannot be decided mutually between them shall be settled under the Rules of Conciliation and Arbitration of the International Chamber of Commerce by one or more arbitrators appointed in accordance with the Rules.

The rules referred to are the Rules of Conciliation and Arbitration, published by the International Chamber of Commerce, who also publish a *Guide to ICC Arbitration*. The ensuing court of arbitration, when formed, is supervised by the ICC itself. Awards by the court are accepted as binding by both parties, as a natural sequence to their agreement to refer disputes to the ICC courts, but awards directly given by the courts cannot, of course, then adopt just selected parts of the English statutes and be enforced as such by the English legal system. They might be enforced by adopting some other cause (such as breach of contract) or stating a case under the law applicable to the contract itself.

8.10 COMMERCIAL COURTS

Finally, a word should be said about a special commercial court of the Queen's Bench Division, which has sat for some years in London with the expressed intention of providing a 'speedy' and 'cut-price' commercial court, devoting its time largely to matters of interpretation of Maritime Law, and the associated subjects of shipping, shipping documents, marine insurance, and the like. It is also able to deal rapidly with other matters of the law as they affect commercial relations between firms. Following some reorganization in 1964, it could prove both quicker and cheaper than straight arbitration for any matters which fall within its scope. For some reason, the commercial court never 'took off' to the extent which had been expected. The commercial court would, of course, lack any special technical knowledge which a duly appointed arbitrator might bring: it is presided over by a judge, and is therefore of most interest to engineering contracts when the root cause of the dispute referred to it is a purely legal one without technical implications. Its services should not be overlooked when a dispute arises.

Appendices 1–14

To aid precision when interpreting a contract, it is usual to begin the conditions of contract with a list of definitions of the terms to be used. Some common words are allotted a narrower, more special meaning than usual. When in the documents the word is intended to have this special meaning, it is spelt with a capital initial letter: when intended to have its normal dictionary meaning, it is used with a lower-case.

The following is a list of typical examples; however, there is no standardization, so that when studying a contract document reference must first be made to the list contained in it. The corresponding terms used in some standard forms of conditions are compared in Appendix 2.

The Employer: The person ordering the goods, works or services covered by the contract, including his legal representatives, assigns or successors.

The Contractor: The person whose offer to provide goods, works or services has been accepted by the Employer, including the contractor's legal representatives, assigns or successors.

The Project Manager: A person named in the contract, or from time to time notified in writing by the Employer to the Contractor, as his manager for the contract or the project of which it forms a part. In default of any such notification, it shall mean the Employer.

The Engineer: The Project Manager in those contracts which name an Engineer and allot duties to him under that title. Always spelt with a capital 'E'.

The Contract: The bargain between the Employer and the Contractor for the provision of the Works, including all those documents to which reference can properly be made in order to ascertain the rights and obligations of the parties to the said bargain.

The Contract Price: The sum to be ascertained and paid in accordance with the contract by the Employer for the execution of the Works by the Contractor.

The Contract Value: The valuation on the basis of the contract price of a stated portion of the Works in the condition and at the place at which the said portion is at the relevant time.

The Works: All plant to be provided and works and services to be done by the Contractor under his contract.

Plant: All machinery, apparatus, articles, materials and things to be provided by the Contractor under the contract, other than Contractor's Equipment.

Permanent Works: All work to be permanently constructed and plant to be supplied connected therewith, for the Employer under the contract.

Temporary Works: All temporary work or plant of every kind required on or about the construction of the permanent works but not for final incorporation therein.

Contractor's Equipment: All tools, tackles, stores, machinery, vehicles, apparatus, articles, materials or things used by the Contractor for the purpose of carrying out his contract, but not for incorporation into the Works.

The Site: The place where plant is to be delivered or works and services to be carried out, together with so much of the surrounding area as the Contractor is allowed to use in connection with the Works, other than purely for the purpose of access to the place.
A Section of the Works: An identifiable portion of the Works which has been so named and described in the contract documents.
Day, Month, and Year: Shall be the same according to the Gregorian calendar (NB: of particular use overseas where they are standardized on something different).
FOB, CIF, C and F, etc.: Shall be as defined in INCOTERMS 1980 and any subsequent revisions or amendments.
Writing: Shall include any manuscript, typewritten or printed statements, under hand or under seal as the case may be. In some contracts telex, fax, computer print-outs, signed by an authorized signatory are accepted as 'writing', but this must be checked. Cables or records of telephone messages are *not* included as 'writing'.
Agreement: An Agreement (spelt with a capital 'A') is a formal substitution or confirmation of a signed contract, introduced subsequently. It does not postpone the contract or starting dates of the existing contract, unless expressly stated in the new Agreement itself.

With reference to the last item, above ('Agreement'), this is often used by customers overseas to mean the *actual contract*, which is frequently drawn up in the form of a legal agreement and signed by the parties. In some cases, this same single document is often called 'the contract', and the only advice which can be given to a project manager is to make certain exactly what is meant in any particular case he might come across. The word Agreement can, of course, also be used for a formal agreement between parties, without any contract actually being involved.

APPENDIX 2 *Equivalent terms used in some standard forms of conditions of contract*

The terms shown in any line have generally the same meaning, but there are cases where the definitions and use are not identical. The terms for FIDIC (Civil) and FIDIC (Electrical and Mechanical) Conditions are the same as those used in the ICE (5th edn) and IMechE/IEE Model Form 'A' Conditions respectively*.

ICE (5th edn)	IMechE/IEE 'A'*	IChemE	Standard Form of Building ('RIBA' or 'JCT')	EEC	BEAMA 'B'	This volume
Employer	Purchaser	Purchaser	Employer	Purchaser	We	Employer
Contractor	Contractor	Contractor	Contractor	Contractor	You	Contractor
Engineer	Engineer	Engineer	Architect/Supervising Officer	–	–	Engineer
(Temporary) Works (Permanent) Works	Works	Works	Works	Works	–	Works
–	Plant	Plant	'materials and goods'	Plant	Goods	Plant
Construction Plant	Contractor's Equipment	Contractor's Equipment	–	Contractor's Equipment	–	Contractor's Equipment
Completion	Completion	Completion	Practical Completion	Completion	Completion	Completion
–	Taking Over	Taking Over	'taking possession of'	Taking Over	Taking Over	Taking Over
'final test'	Tests on Completion	Take-Over Tests	–	Take-Over Tests	–	–
Certificate of Completion	Taking Over Certificate	(Taking Over Certificate) Acceptance Certificate	Certificate of Practical Completion	Take Over Certificate	–	Taking Over Certificate
'a certificate'	Interim Certificate	'a certificate'	Interim Certificate	–	–	–
Maintenance Certificate	Final Certificate	Final Certificate	Final Certificate	–	–	Final Certificate
Period of Maintenance	(No specific term)	Defects Liability Period	Defects Liability Period	Guarantee period	–	Period of Warranty
Contract Price	Contract Price	Contract Price	Contract Sum	–	Contract Price	Contract Price
–	Contract Value	–	–	–	Contract Value	Contract Value

* See also pp. 173–4, A (iii) and C (i).

In the engineering field one can expect to come across contracts with the following titles; it will be seen that they can be classified in either of two ways: the basis of their negotiation of terms, or the objective with which they have been formed.

Classification by method of negotiation

• Fixed-Price Contract.	The contract price is settled from the outset. The employer knows exactly what he will have to pay. The contractor takes all the risks (including inflation and under-estimates); the Employer can expect to pay a big price for his immunity!
• Lump-Sum Contract.	The contract works are considered as a single item and an all-inclusive price is named. CPA can be applied if desired.
• Bills of Quantity Contract, Schedule of Rates Contract.	The contract works are subdivided as finely as desired and are priced item-by-item. Standard methods of listing and measurement are named.
• Cost-plus Contract.	The contractor is paid his total costs plus an agreed price for his overheads and for his profit. The latter can be a 'fixed fee', a 'limited fee' (i.e. a stated maximum), or a 'percentage fee' (the usual). Used mostly for development work or when a final design cannot be foreseen.
• Target-Cost Contract.	An attempt to combine 'Cost-Plus' and 'Competitive Tendering'. If there are many tenderers, they each name their 'target price'. If only one, he may be given it by the Employer. In either case, they get only an agreed percentage of any excess over their agreed target, but a bonus for any reduction in price attained. The problem for a project manager is to establish a 'fair' target price.
• Competitive Contract.	Contractor chosen by competition, usually on his price and delivery, for a specified contract works.
• Negotiated Contract.	A contract finally agreed with a single contractor, both sides giving way, in the process of negotiation from an original offer.

Classification by the objective of the contract

• Package Contract.	Two or more related contracts: take all or none! Value to Employer can be: (a) technical continuity, (b) fewer contractors on-site and (c) inducement to a contractor to take one unattractive job.
• Turnkey Contract.	A package contract involving several engineering disciplines under one main contractor. Hands over complete operative project to Employer.
• Running Contract.	Exclusive supply of goods or services at specified intervals or over a stated period of time. Often against an estimate of total demand or a guaranteed minimum total.
• Service Contract.	Contract for services only – e.g. research, advice, drawings, computer programs, etc. Also used for supply of services – e.g. gas, water, phones, electricity, etc.
• 'Continuation', 'Serial' or 'Extension' Contract.	Contract for immediate work, which also makes provision for future contracts for extensions. Could use same basic contract terms. Permits rapid switch of plant, manpower, machinery, etc. to later jobs. The same term is also used for the subsequent contracts themselves when so associated.

Most forms of contract can be at fixed prices or could involve Contract Price Adjustment clauses if so agreed. Exception is, of course, a fixed-price contract. We consider CPA at some length in Volume 2, *Competitive Tendering.*

The following are some of the terms met with commercially in engineering contract work, and which are recognized in commercial transactions. They may not appear in the present set of volumes.

Abandonment:	An insurance term: the insurer takes over damaged goods in return for settlement of total loss by insurer.
Acceptance (of a Bill of Exchange):	Signature of a Bill of Exchange by the buyer to signify he agrees the payment of the sum shown and the date of maturity.
Accommodation Bills:	Bills of Exchange which have been drawn to raise funds or credit, and not to pay a debt.
Account Stated:	Acceptance by two parties of a combined statement of their indebtedness to one another.
Ad Valorem:	'According to its value.' Often applied to a rate of Import duty, etc.
Adjudication Order:	An order by a Bankruptcy Court to the insolvent party to accept its nominated manager or trustee in bankruptcy.
Air Consignment Note:	A receipt given to the seller or his agent when goods are deposited for transportation by air freight. Equivalent to a Bill of Lading but for air carriage.
Allonge:	Extension sheet for additional endorsements to a Bill of Exchange.
Barratry:	A wrongful act by the master or crew of a ship, which affects its owner or its charter.
Bill of Exchange:	A negotiable instrument of indebtedness which matures at the agreed date shown on it.
Bill of Lading:	A formal receipt by the Master of a ship for goods taken on board from the seller for carriage by sea to a specific port.
Bull:	A person who buys stocks and shares which he hopes to sell at a higher price before he has to pay for them himself (at the next 'settlement day').
Bullion:	Gold or Silver in some form other than coins or medals (usually in bar-form, but not necessarily).
Caveat Emptor:	'Let the buyer beware.' Applies to any contract in which the buyer uses his own judgement about what he buys (and not the seller's proposals).
Champerty:	Continuing a court case on the understanding that the person concerned acquires a stated part of the award (damages) given.
Collateral Security:	Security given to a lender by a borrower to back a loan he is to be given.

Currency (of a Bill of Exchange):	The time which has to run before the Bill of Exchange matures.
Days of Grace:	Three days, allowed after the maturity date of a Bill of Exchange for payment to be made.
Domicile:	The country or place where a person lives or a firm has its registered offices.
Drawback:	Repayable import duty, when imported goods are re-exported or proved to the customs authorities to have been destroyed.
ECGD:	Export Credit Guarantee Department. A governmental office which insures exporters (on payment of assessed premiums) against non-payment of purchase prices due to buyer's insolvency, political instructions, economic difficulties, etc. The ECGD must approve any contract of sale beforehand.
Escrow:	. . . to be in . . . A Deed which does not come into effect until a prior condition has been fulfilled.
Estoppel:	Prevention of a person denying a previous statement which induced another party to take an action he would not otherwise have done.
Flotsam:	Floating goods from wrecked ships.
Foreclosure:	Seizure of security deposited as the collateral for a loan.
Freight:	A shipowner's charge for the safe carriage of a specified cargo in his ships.
Goods (Ascertained):	Goods already made or existing.
Jetsam:	Cargo or part of a ship thrown overboard to lighten the vessel and save the remainder.
Jettison:	The act of causing Jetsam.
Liens:	The rights of a person to retain possession of things, or to have things not in his possession, unless and until some liability is settled.
Man of Straw:	A person not worthy of being given credit.
Manifest:	A ship's document giving details of crew and cargo sent by the Master to his owner or his owner's representative before sailing. A similar document for air freight.
Mate's Receipt:	A temporary receipt given to the exporter for goods loaded aboard, until such time as a formal Bill of Lading can be prepared. The 'mate' is the ship's officer in charge of loading aboard.
Parol Contract:	A simple contract (i.e. under hand).
Salvage:	The act (or its reward) by a third party voluntarily saving property at sea.

Sets of Bills:	Bills of Exchange are customarily produced in small sets, sent or kept separate for security, of which only one copy is designated to be redeemed at maturity. The rest become void and are destroyed.
Sight Draft:	Any money draft (cheque, etc.) which is dated and marked 'on sight' (i.e. payable on sight). Travellers cheques, etc. are all 'sight drafts' payable on presentation.
Stoppage 'in Transitu':	Retention of the goods by the carrier at the port of destination when the buyer has not paid for them. Done by marking Consignee 'to order' on the Bill of Lading until the exporter releases the goods on payment being received.
Subrogation:	An insurance term. After settling for a loss, the insurers are entitled to the compensation made by any other source, up to the sum they have paid out.
Uberrima Fides:	'Utmost good faith.' Contracts (usually insurance) which require a full disclosure of all relevant facts, not just an absence or passing reference, or an abbreviation.
Ultra Vires:	Beyond one's legal authority.
Void:	A contract of no legal effect.
Voidable:	A contract capable of being disclaimed in part or as a whole at the option of one of the parties thereto.

APPENDIX 5 *Terms used in overseas transportation of engineering goods*

Goods sent from a UK seller to a buyer in an overseas country

Ex-Works:	Dispatched ex-factory or warehouse, off the ground, on to Buyer's vehicles.
FOR, FOT:	Free on to rail cars or on to trucks at the Seller's factory or warehouse, or at named loading point.
FAS:	Free alongside ship at named port and quay
FOB:	Free on Board ship at named port. The same as FAS, plus cost of moving goods across the ship's rail. Buyer arranges shipment.
C&F:	Cost & Freight. FOB plus cost of carriage to named port of destination, but still on board ship. Seller arranges shipment.
CIF:	Same as C&F, plus cost of marine insurance to named port of destination.
Ex Ship:	Unloaded from ship and put at disposal of Buyer alongside, at port of destination.
Ex Quay:	Ex-ship, but import duty paid at port of destination by Seller.
Delivered at Named Frontier:	Does not include payment of import duty to Buyer's country.
Delivered to Site:	To Buyer's site with all costs (including customs and other dues) and cross-country transport paid by Seller.
Delivered to Site (TIR)	To Buyer's site by Seller's transport. Bodywork of transport has to be approved. Goods are checked on to vehicle and sealed at the Seller's premises by home customs official. Unsealed and checked off against invoices by customs official of importing country at Buyer's site. Buyer then pays his import duties and any other service payments at receiving end (including unloading costs).
FOB Airport:	Same as FOB, but rarely will Seller be allowed to load on to plane. Delivery to carrier's warehouse at airport, with costs of loading on to aircraft included.
Bill of Lading:	Shipmaster's formal receipt, etc. for goods taken on board. Certificate of 'shipment'.
Air Consignment Note:	Equivalent of Bill of Lading, when air freight is used instead of ship.
Shipping Documents:	Usually include: Bill of Lading, Invoices, packing lists, Bill of Exchange (if not paid for separately) or receipt for cash due, insurance certificate or policy. Any other document asked for by Buyer (e.g. export licence or import licence).

APPENDIX 6 *Typical sets of Standard (or Model) Conditions of Contract found in engineering projects*

The terms of a contract must deal with those eventualities normally expected to occur during its execution, which affect the interests of one or both of the parties. It is better to settle these during the negotiation of the contract, rather than after they have occurred and the two parties are in the heat of an argument. To save the contract draughtsmen having to think out and include all such clauses every time they make a contract a number of bodies, both professional and commercial, have drawn up model sets of clauses, which usage and experience have shown deal satisfactorily with the foreseen eventualities when they arise.

The model clauses are published in sets and are designed to be used with most engineering contracts. Although their clauses may be to some extent interdependent and interlocking, the sets are not expected to be exclusive, and special clauses can be added by competent draughtsmen to meet the particular circumstances of individual contracts.

The following list gives most of the standard sets of conditions commonly met with in the engineering industry in the UK. It is not intended to be complete, nor to be set out in any order of priority, comprehensiveness or frequency of use. The items shown are, in many cases, only a few of the reference documents published by the different organizations mentioned, but they include those most commonly met with by project managers.

A. Issued by professional institutions

(i) Institution of Civil Engineers (ICE)
  - (a) General Conditions of Contract for Works of Civil Engineering Construction (5th edn).
  - (b) General Conditions of Contract for Minor Works (1988 edn).
  - (c) Conditions of Contract for Ground Investigations (1983 edn).
  - (d) Form of Subcontract.

(ii) Institution of Chemical Engineers
  - (a) Model Form of General Conditions of Contract (Process Plants) (not export).
  - (b) Model Form of General Conditions of Contract (Process Plants – Cost-plus Contracts) (not export).

(iii) Joint Committee IMechE/IEE/ACE

(a) Model Form A
(b) Model Form B3 } Superseded by Model Form 1 in June 1988.

(c) Model Form B1
(d) Model Form B2 } Similarly, to be superseded in the near future.

(e) Model Form C
(f) Model Form E } (Cabling.) Recently revised: no replacement plans.

(iv) Institute of Purchasing and Supply

Model Forms Nos 1–10 cover a number of specialized and general subjects and are available in bound sets in book form; they include:

(a) Model Forms 1 and 2: For general engineering plant or materials.

(b) Model Form 4: For Minor Works or for Engineering Services.

(c) Model Form 8: For repair or modification of engineering equipment.

There is a further form (hitherto unnumbered) which is designed for Computer Equipment (1987).

(v) Joint Contracts Tribunal (RIBA, etc.)

(a) Standard Form of Building Contract (6 versions).

(b) Standard Form of Building Contract with Contractor's Design (1981).

(c) Standard Form of Contract for Nominated Subcontractors (6 versions).

(d) Standard Agreement for Minor Building Works (IFC 84).

(e) Standard Agreement – Employer/Nominated Subcontractor.

(f) Standard Form of Contract – Management Contract.

(vi) British Electrical and Associated Manufacturers (BEAMA)

General Conditions of Contract (many versions, some for particular electrical products – e.g. electronic equipment). Although intended primarily for electrical products, these conditions of contract are of equal application in general cases of supply of smaller machinery.

B. Major organizations and users

(i) Electricity Board/BEAMA

Conditions A (Plant with erection).

Conditions B (Plant without erection).

(ii) HM Government

Works Contracts (GC/Works/1).

Minor Works Contracts (GC/Works/2) (1980).

Works Contracts – Stores (GC/Stores/1).

Works at Government Establishments.

Building, Civil Engineering, Mechanical and Electrical Small Works (C. 1001–1985).

(iii) UK Atomic Energy Authority

General conditions for Building Works.

Standard Conditions for Stores Purchases.

(iv) British Steel Corporation
General Conditions of Contract.
There are many other similar conditions of contract by big users.

C. For international use

(i) FIDIC (Fédération Internationale Des Ingénieurs-Conseils)
(a) Conditions of Contract for Works of Civil Engineering Construction (4th edn, 1987).*
(b) Conditions of Contract for Electrical and Mechanical Works including Erection (3rd edn, 1987).*
(c) Project Management Contracts – IGRA (1980) PM.
(d) Numerous other leaflets and guidelines.
(e) International Model Form of Agreement – Consultant.

(ii) United Nations – ECE (United Nations Economic Commission for Europe)
General Conditions for Supply of Plant and Machinery, etc. (Form 188).
General Conditions for Supply of Plant and Machinery, etc. with Erection (Form 188A).
Additional Clauses for Supervision of Erection (Form 188B).
General Conditions for the Erection of Plant Machinery, etc. (Form 188D).

* The main conditions of FIDIC are issued as Part I (the general conditions) and Part II (conditions for particular applications).

APPENDIX 7 *Typical contents of the Major Model Conditions of Contract used in the engineering industry*

The list below gives the subjects which can be expected to be covered by most of the Major Model Conditions of Contract (see Appendix 6). No two models are the same, however, and more especially, those designed for use with the provisioning of plant (where details of design are left to the firm in whose factory the plant is manufactured) will differ in many details from models designed for use with civil engineering contracts (in which details of design are determined by the Employer). Subjects like re-measurement, or those dealing with nominated subcontracting, will not be mentioned in the 'plant' category of model conditions.

Each standard is designed to have wide application in its own field and is correspondingly general in its terms. It will often be necessary to add special conditions when dealing with individual contracts, supplementing or modifying the more general clauses, even in some cases deleting them altogether – terms of payment may be a case in point. In most cases, some clauses will call for the addition of a figure (for example, the rate of liquidated damages for delay), where blank spaces are left for this purpose in the model.

The order in which different models treat particular subjects will be different too; in what follows they are listed in roughly alphabetical order (but allied subjects may be taken together).

Arbitration and appointment of arbitrators.
Arbitration and appointment of arbitrators, continuance of work during.
Assignment of contract and subletting.
Boreholes and underground works.
Completion dates, for separate sections of the work.
Contract documents, lists of.
Delivery dates.
Contract prices, terms of payment.
Contract Price Adjustment, base dates, methods of.
Contractor, nomination of, supervision by, on site.
Contractor, death and bankruptcy, insolvency.
Subcontractors, approval by Employer, nominated subcontractors, preferred subcontractors.
Costs, prime and provisional.
Disputes, treatment of, liability for damages.
Definitions.
Delay, damages for; sections of the Works.
Facilities on site for contractors.

The Engineer, nomination and responsibilities of.
Design, responsibility for, liaison on.
Final Certificate, contents of.
Fair Wages Clause, government call for.
Fossils, antiques and treasure, care and recovery of.
Guarantees, Bonds, etc. by contractors, Agreements.
Inspection and testing by Employer.
Information from Employer, delay or mistakes in.
Nominated Subcontracting, position of main contractor.
Lighting and power requirements on site.
Licences to use patented procedures.
Lifting equipment provided by Employer on-site.
Packing and transport for delivery to site.
Patents and patent liabilities.
Property in, and Vesting of, contractor's plant at site.
Measurement, methods of, re-measurement, rules for.
Payments of contract price, times of, methods.
Payments of contract price, currency of.
Rejection of work, Engineer's decision.
Responsibility of contractor for works.
Workforce, machinery, employees, supervision at site.
Safety and Security measures at site: Health and Safety at Work, etc. Act 1974, duties under.
Site Regulations.
Storage at site.
Site access.
Suspension of Works by Employer.
Tests to be made by Employer, certification of.
Taking Over, certification.
Setting out of works, responsibility for.
Representative, the Engineer's nomination of.
Variations of Specification of the works, limits on.
Valuation of variations authorized.
Workmanship and materials quality, Employer to inspect and approve or reject.
Work covered up after inspection.
Warranty clause, period of warranty, repair work.
Insurances by contractor, Employer to approve policies, indemnity and payment of premiums.
Damages, loss or injuries sustained by Employer or third parties.
Statutory or Local Regulations to be observed.

Site clearance on completion of contract work.
Day-work only after authorization.

Note: Project manager to be read as alternative to the Engineer throughout the above.

Section 266 of the Local Government Act 1933 provides as follows. It is on this section that a Contractor relies when he accepts a contract with a permanent official of local government, acting as Buyer's representative:

Section 266 (1) A local authority may enter into any contracts necessary for the discharge of any of their functions.

(2) All Contracts made by a local authority or by a committee thereof shall be made in accordance with the Standing Orders of the local authority, and in the case of contracts for the supply of goods or of materials, or for the execution of works, the Standing Orders shall require that, except as otherwise provided by or under the Standing Orders, notice of the intention of the authority or committee, as the case may be, to enter into the contract shall be published and tenders invited. The manner in which such notice shall be published and tenders invited shall be regulated by the authority's Standing Orders.

Provided that: A person entering into a contract with a local authority shall not be bound to enquire whether the Standing Orders of the authority which apply to the contract have been complied with, and all contracts entered into by a local authority shall, if otherwise valid, have full force and effect notwithstanding that the Standing Orders applicable thereto have not been complied with.

By dint of constant usage, the following wording has become standardized in the engineering industry, for a declaration of a tender being bona-fide. It has no other formal authority.

Declaration of bona-fide tender

The essence of selective tendering is that a Purchaser shall receive bona-fide competitive tenders from all firms tendering.

In recognition of this principle, we certify that this is a bona-fide tender, intended to be competitive, and that we have not fixed or adjusted the amount of the tender by or under or in accordance with any agreement or arrangement with any other person. We also certify that we have not done, and we undertake that we will not do at any time before the returnable date for this tender, any of the following, that is to say:

(a) communicate to a person (other than the person calling for these tenders) the amount or approximate amount of the proposed tender;
(b) enter into any agreement or arrangement with any other person that he shall refrain from tendering or as to the amount of any tender he will submit;
(c) offer or pay or give or agree to pay or give any sum of money or valuable consideration directly or indirectly to any person for doing, or having done, or causing, or having caused to be done, in relation to any other tender or proposed tender for the same work, any act or thing of this sort, as described above.

In this certificate the word 'person' shall include any person and any body or association, corporate or incorporate; and 'any agreement or arrangement' shall include any such transaction, formal or informal, and whether legally binding or not.

SIGNED (as in the accompanying Tender):

....

On behalf of: ....

DATE: ....

To whom it may concern

At their Ordinary meeting held at .... on ....
by a unanimous vote of those present, the Directors of this Company comprising its Board of Management adopted and passed the following Resolution. I am instructed by them to send you a copy and I hereby certify that the following is a true copy of Resolution No. XX/90 as contained in Minute No. YY of the 1990 normal proceedings of the Board:

RESOLUTION No. XX/90
It was RESOLVED and APPROVED that .... [name in full] .... of .... [address] ....,
a member of [this organization] be assigned and empowered to use in the name of, and on behalf of [this organization] in connection with ....
.... [state purpose for which the Warrant is issued] ....
all such acts, negotiations and approvals as he shall think fit, without further reference to us, provided always that:

All decisions shall relate to the activity specified herein.

This warrant shall become null and void on [date] unless formally extended by us in writing.

Such acts, etc. shall not (without prior authority from us in each and every case) be such as to involve [this organization] in any undertaking which shall exceed a total commitment of more than £.................... sterling or equivalent in other currencies.

Signed and certified: ....

Date: ....

Place:...............................   ....
(Secretary to the Board)

APPENDIX 11 *Typical collateral agreement between employer and intended nominated subcontractor*

This form of Agreement can be used with a contractor whom the Employer expects to nominate but who has meanwhile signed a contract with the Employer. The terms of this contract will, it is hoped, eventually be transferred to a nominated subcontract with the Main Contractor.

THIS AGREEMENT made this .... day of ....
19........ between ....
of ....
(hereinafter called 'the Employer') of the first part and ....
.... of ....
.... (hereinafter called 'the Subcontractor')
of the second part now WITNESSETH that:

WHEREAS the Employer intends entering a contract (the 'Main Contract') with a contractor (the 'Main Contractor') for ....
.... (the 'Works') subject to terms and Conditions yet to be negotiated but generally in accordance with ....
....
....

AND WHEREAS the Subcontractor has been invited by the Employer to tender and has so tendered ('the Offer') for ....
.... (the 'Subcontract Works') forming a part of the Works, and the Employer will nominate the Subcontractor as a Nominated Subcontractor to the Main Contractor (when appointed) in accordance with the provisions of clauses 58(5) and 59A of the standard conditions of contract published by the Institution of Civil Engineers for use with works of civil engineering construction (5th edition) it is NOW agreed between the parties as follows:

1. In consideration of the Employer having invited the Subcontractor to tender for the subcontract works, and having duly accepted the Offer, and it being proposed that the Subcontractor shall become a Nominated Subcontractor to the Main Contractor (when appointed) the Subcontractor by this agreement undertakes

(a) to hold and maintain the Offer contained in his tender and accepted by the Employer for a period of .... weeks from the date of signing of this Agreement, and
(b) to transfer the Offer in the same terms and conditions *mutatis mutandis* for acceptance by the Main Contractor, subject only to such minor adjustments (if any) as may be proposed by the Main Contractor after appointment, and
(c) to enter into a nominated subcontract with the Main Contractor on the agreed basis of the Offer to carry out the Subcontract Works, and

(d) As and when such nominated subcontract is established between the Subcontractor and the Main Contractor, the above contract with the Employer shall automatically without further action by either party become null and void.

2. This Agreement shall likewise become null and void and determine if either

(a) the period specified in Article 1(a) above or of any extension thereof subsequently agreed between the parties hereto shall have elapsed, or
(b) The Employer shall at any time give notice to the Subcontractor in writing of his intention to determine the Agreement by reason of either

(i) the said Offer having proved unacceptable to the Main Contractor; or
(ii) the Main Contractor in the exercise of his rights under the terms of the Main Contract having validly objected to the nomination of the Subcontractor; or
(iii) the Employer for any reason whatever having elected not to proceed with the Works or with the part of the Works which contains the Subcontract Works.

3. Termination of this Agreement in accordance with Article 2 shall be without liability to either party. Any charges legally incurred by the Subcontractor in connection with the execution of the Subcontract Works shall upon nomination be chargeable against the Main Contractor, and recoverable by him against the prime-sums included in the Main Contract for that purpose.

4. Nothing contained in the Offer or in any modifications thereto duly negotiated and agreed between the parties hereto shall operate to limit the Subcontractor's obligations under the terms of this Agreement.

IN WITNESS WHEREOF the parties hereto have set their hands through their duly authorized representatives and before witnesses, the date first above mentioned.

SIGNED for and on behalf of:	SIGNED for and on behalf of:
....	....
....	....
by:	by: ....

Witnesses to the above signatures:

Name	Name ....
Occupation	Occupation ....
Signature	Signature ....

Guarantee of due performance

BY THIS BOND made this .... day of .... one thousand nine hundred and ...., WE .... (hereinafter called 'the Contractor') and .... (hereinafter called 'the Sureties') are held and firmly bound unto .... (hereinafter called 'the Employer') so that

WHEREAS by an Agreement (hereinafter called 'the Agreement') dated .... the Contractor and the Employer entered into an Agreement as aforesaid

NOW WE , the Sureties hereby jointly and severally guarantee to the Employer punctual true and faithful performance and observance by the Contractor of the covenant on the part of the Contractor contained in the Agreement and undertake to be responsible to the Employer his legal personal representatives successors or assigns as Sureties for the Contractor for the payment by the Contractor of all sums of money, losses, damages, costs, charges and expenses that may become due or payable to the Employer his legal personal representatives, successors or assigns by or from the Contractor by reason of or in consequence of the default of the Contractor in the performance or observance of the said covenant on the part of the Contractor but so nevertheless that the total amount to be demanded or recovered by the Employer his legal representatives, successors or assigns of or from us as Sureties shall not exceed .... pounds sterling (£.....................)

This Guarantee shall not be revocable by notice and our liability as Sureties hereunder shall not be impaired or discharged by any extensions of time or variations or alterations made, given, conceded or agreed (with or without our knowledge or consent) under the terms and conditions contained in the Agreement or (where the Employer or Contractor is a Firm) by a change in the constitution of the Employer's or the Contractor's Firms

Demands for payment under this Guarantee shall be made upon us in writing and such payment by us shall be made without objection on our being given evidence as to the existence of a default by the Contractor and of the damages due or payable to the Employer in respect thereof which shall be a certificate of the findings of a court of law or of arbitration (in cases in which the default has been the subject of proceedings therein) or otherwise the certificate of the person named as the Project Manager or as the Engineer in the Agreement aforesaid

IN WITNESS WHEREOF the Contractor and the Sureties have caused their respective Common Seals to be affixed hereto, the day and year first before written

(Common Seals of the Contractor and the Sureties fixed and witnessed here)

APPENDIX 13 *Typical Instruction to Proceed*

(Sent under Employer's letterhead)

(Tenderer's name and address): .... Reference: ....

.... Date: ....

....

Dear Sirs,

[Project]
....

Tender No.:............................ for ....[Contract title].....

1. We are pleased to inform you that, subject to a satisfactory outcome to negotiations between us on certain points in your above-mentioned tender, the contract for this work will be placed with your company.

To allow work to be put in hand without delay will you kindly accept this letter as an INSTRUCTION TO PROCEED subject to the terms and conditions of contract other than those in dispute between us, with the following parts of the said contract:

(a) ....
(b) ....
(c) ....

2. This instruction is subject to an overriding limit of total expenditure by you (incurred and committed) of a contract value of .... pounds sterling (£................) which figure shall not be exceeded without written authority from me.

3. On placing the full contract with you, all liabilities and benefits of both parties will be absorbed thereby and this INSTRUCTION TO PROCEED shall be determined. In the event the negotiations aforementioned fail to lead to agreement so that a full contract cannot be concluded between us, we shall have the right to determine by notice in writing this INSTRUCTION TO PROCEED and you will be paid the contract value of all contract works you shall have properly executed, together with such other costs and expenses as you have necessarily incurred as a result of this instruction, including any partly finished works. Property in all materials acquired and work carried out shall forthwith vest in us (if not already so vested under valid contract terms) and be held available by you at our disposal.

4. Will you please let me have (by first-class return mail) your unconditional acceptance of this INSTRUCTION TO PROCEED, and thereupon put the specified work in hand without delay.

Yours faithfully,

(Signature):...

(Position): ....

FOR AND ON BEHALF OF:

[Employer]
....

APPENDIX 14 *The Copyright, Designs and Patents Act 1988: a brief synopsis*

The draft of this far-reaching Act has been considerably changed following approaches to the government by many interested groups. Its final form has now been passed by Parliament, when it was stated by the Minister concerned: 'its effects would be felt to a greater or lesser degree by almost every sector of the economy.' It is a long Act, bringing up to date all earlier legislation on the subjects and adding new rules to meet the needs of the modern electronic age. In so doing, it amends some of the details given elsewhere in this book (notably Section 5.10). It comes into effect during 1989 but meanwhile only a brief synopsis of its precise limits of application can be included.

The Act is far from precise in many of its aspects and will take a number of years and considerable case law before it can be set out authoritatively. It also brings new concepts into English law, not recognized hitherto. Broadly speaking, it extends present protection to some extent, but it also recognizes modern electronic processes such as, for example, computers, telecommunications, digital systems, etc.

SOFTWARE (not precisely defined) is given greater protection, including not only computer programs themselves, but tables, compilations, and the like, as well as specifications or manuals of instruction for computers. All are now to be regarded as 'literary works' and are automatically protected as soon as they are created. Protection is now given for a period of the lifetime of the creator, plus a further 50 years. If the creator is a paid employee of a firm, or if he is working under an arrangement which expressly assigns his copyright, the firm or employer concerned is automatically given the rights accruing to the creator. Such rights now cover the importation of infringing copies, unauthorized changes to the work (such as translation, etc.), and the case of designs or work generated by a computer alone (i.e. when no human author is concerned). The use of a pure computer-operator (who does not himself actually 'create' the work in question) and the composer of the program which enables the computer to do so unaided are not dealt with: any 'rights' are 'owned' by 'the person making the necessary arrangements' (whatever this might mean – the owner of the computer perhaps?). This, then, is one of the many points the courts will have to define.

All software to be protected must be 'recorded' (i.e. in writing, printing or in some form of notation or code). It can be in any medium (undefined, but it can be broadly interpreted as 'however recorded', and could include magnetic recording on a coated tape). The making of transient copies, storage, reception over a telecommunication circuit, all forms of distribution with, say, computer installations, or as scientific abstracts might be considered illegal unless licensed by the copyright-owner – but see the reference to a proposed licensing scheme below.

A new concept of UN-REGISTERED DESIGN is introduced for any original articles (i.e. hardware, not literary articles). They are to be given automatic protection as an 'unregistered design' for a period of 10 years: during the second 5-year period the owner of the protective rights must grant licences on request at rates of royalty he (and the courts, if applied to) regard as 'reasonable'. There are many exceptions under the Act:

- The design and/or function must be original: its design cannot be 'commonplace'.
- The principles of construction may not be widely known.
- It does not apply to components or piece-parts (however unique) designed purely to fit into a larger article; hence it cannot apply to spare-parts.

Registered Design rights continue, and articles must have some relevance to a purchaser (customer) before some design can be accepted for 'registration'. A Registered Design has a protection period of 25 years.

ENGINEERING DRAWINGS are still copyright and protected during the lifetime of the draughtsman, plus a further 50 years. Oddly enough, an article produced from the drawings (assuming it is 'original' in concept) may be protected for 10 years (as an 'unregistered design') only; a competitor can thereafter copy the product, even though he must still produce his own (non-infringing) drawings! There must inevitably be some resemblance between his own drawings and the originals as the same article is produced from both.

The law regarding multiple copying by schools, information services, research establishments, etc., and the copying by individuals for study, are made more precise, and the Act envisages the setting up of a licensing scheme for such purposes, with suitable royalties to be agreed beforehand. There are also further new rules relating, for example, to professionals, such as consultants or experts, who have contributed to material assembled and protected by somebody else. Otherwise a professional might give his plans to the world at large without himself having any protection. His views might be misconstrued to his detriment.

The concept of MORAL RIGHTS (already used abroad as *Droits morales*) is now introduced into English law, and an associated person can be given a measure of protection against his clients who have used his expert knowledge when creating their own original works.

Copies of the Act itself can be obtained from the HM Stationery Office in the usual way, but intending readers are warned that it is both long and complicated and will need study by experts. Even then a lot is still undetermined by case law, especially in regard to how far its regulations can be considered to apply.

Index

Entries in *italics* refer to the Appendices.

For Product Safety Concerns and Information please contact our EU representative GPSR@taylorandfrancis.com
Taylor & Francis Verlag GmbH, Kaufingerstraße 24, 80331 München, Germany

www.ingramcontent.com/pod-product-compliance
Ingram Content Group UK Ltd.
Pitfield, Milton Keynes, MK11 3LW, UK
UKHW021959180926
475701UK00003B/102
* 9 7 8 0 3 6 7 5 8 0 2 0 9 *